**Daniel Owen Spence** is a postdoctoral research fellow in the Centre for Africa Studies at the University of the Free State. He obtained his PhD from Sheffield Hallam University in 2012, holds an Innovation Scholarship with the National Research Foundation of South Africa, and is a fellow of Leiden University's African Studies Centre Community. With research conducted in over a dozen countries, he has published articles and chapters in several peer-reviewed international journals and edited collections, and is the author of *Colonial naval culture and British imperialism, 1922–67* (2015).

*A History of the Royal Navy: The Age of Sail*
Andrew Baines (ISBN: 978 1 78076 992 9)

*The Royal Navy: A History Since 1900*
Duncan Redford and Philip D. Grove (ISBN: 978 1 78076 782 6)

*A History of the Royal Navy: Air Power and British Naval Aviation*
Ben Jones (ISBN: 978 1 78076 993 6)

*A History of the Royal Navy: The American Revolutionary War*
Martin Robson (ISBN: 978 1 78076 994 3)

*A History of the Royal Navy: Empire and Imperialism*
Daniel Owen Spence (ISBN: 978 1 78076 543 3)

*A History of the Royal Navy: The Napoleonic Wars*
Martin Robson (ISBN 978 1 78076 544 0)

*A History of the Royal Navy: The Nuclear Age*
Philip D. Grove (ISBN: 978 1 78076 995 0)

*A History of the Royal Navy: The Royal Marines*
Britt Zerbe (ISBN: 978 1 78076 765 9)

*A History of the Royal Navy: The Seven Years' War*
Martin Robson (ISBN: 978 1 78076 545 7)

*A History of the Royal Navy: The Submarine*
Duncan Redford (ISBN: 978 1 78076 546 4)

*A History of the Royal Navy: The Victorian Age*
Andrew Baines (ISBN: 978 1 78076 749 9)

*A History of the Royal Navy: Women and the Royal Navy*
Jo Stanley (ISBN: 978 1 78076 756 7)

*A History of the Royal Navy: World War I*
Mike Farquharson-Roberts (ISBN: 978 1 78076 838 0)

*A History of the Royal Navy: World War II*
Duncan Redford (ISBN: 978 1 78076 546 4)

# A HISTORY OF THE
# ROYAL NAVY

## Empire and Imperialism

Daniel Owen Spence

*in association with*

'The fortunes of the British Empire and the Royal Navy were inextricably linked, the one entirely dependent on the other. But the navy was much more than just a fighting force, designed to defeat rivals and subdue subordinate peoples, for it performed many other functions. Its personnel were key in the pursuit of diplomatic ends, in the conveyance of leading political, administrative and royal travellers, in the protection of trade, in the development of both exploration and scientific research, in developing such policies as the export of convicts, in surveying and hydrographic work, in the development of marine, navigational and armaments technologies, and in simply "flying the flag". As this concise yet wide-ranging and comprehensive book demonstrates, the navy also performed a vital psychological role, often on a global scale. It was celebrated in all sorts of visual, literary and cultural forms. In the twentieth century, its significance in both warfare and in relationships with indigenous peoples changed dramatically, and the decline of empire and of naval strength occurred in parallel. Few authors have described these connections more successfully than Daniel Owen Spence.' – **John M. MacKenzie, Emeritus Professor of History, Lancaster University**

'This jaunty and comprehensive account of the navy's association with Britain's maritime empire contributes to a fine tradition of imperial and naval history-writing. A model of compression, sweeping yet with an eye for arresting detail, it captures the spirit of Britain's engagement with the wider world and the navy's intimate role in it. Daniel Owen Spence reveals how the Royal Navy was central to the rise of imperial Britain, its history entwined with commerce, culture, religion, science, exploration, and the accretion of knowledge and power to the British state. At every stage of the process, the navy was affecting the lives of people overseas, and recruiting them into its ranks.' – **Ashley Jackson, Professor of Imperial and Military History, King's College London**

*For Auntie Lel and Uncle Geoff, who helped fire my fascination with the Royal Navy and the sea*

First published in 2015 by I.B.Tauris & Co. Ltd
Reprinted in 2017
London • New York
www.ibtauris.com

ISBN: 978 1 78076 543 3
eISBN: 978 0 85773 961 2

A full CIP record for this book is available from the British Library
A full CIP record is available from the Library of Congress
Library of Congress Catalog Card Number: available

Printed and bound by CPI Group (UK) Ltd, Croydon, CR0 4YY

# Contents

# List of Illustrations

## Tables

## Figures

## Colour Plates

# Series Foreword

The Royal Navy has for centuries played a vital if sometimes misunderstood or even at times unsung part in Britain's history. Often it has been the principal – sometimes the only – means of defending British interests around the world. In peacetime the Royal Navy carries out a multitude of tasks as part of government policy – showing the flag, or naval diplomacy as it is now often called. In wartime, as the senior service of Britain's armed forces, the navy has taken the war to the enemy, by battle, by economic blockade or by attacking hostile territory from the sea. Adversaries have changed over the centuries. Old rivals have become today's alliance partners; the types of ship, the weapons within them and the technology – the 'how' of naval combat – have also changed. But fundamentally what the navy does has not changed. It exists to serve Britain's government and its people, to protect them and their interests wherever they might be threatened in the world.

This series, through the numerous individual books within it, throws new light on almost every aspect of Britain's Royal Navy: its ships, its people, the technology, the wars and peacetime operations too, from the birth of the modern navy following the restoration of Charles II to the throne in the late seventeenth century to the war on terror in the early twenty-first century.

The series consists of three chronologically themed books covering the sailing navy from the 1660s until 1815, the navy in the nineteenth century from the end of the Napoleonic Wars, and the navy since 1900. These are complemented by a number of slightly shorter books which examine the navy's part in particular wars, such as the Seven Years' War, the American Revolution, the Napoleonic Wars, World War I, World War II and the Cold War, or particular aspects of the service: the navy and empire, the Women's Royal Naval Service, the Royal Marines, naval aviation and the submarine service. The books are standalone works in their own right, but when taken as a series present the most comprehensive and readable history of the Royal Navy.

**Duncan Redford**
**NMRN**

The role in Britain's history of the Royal Navy is all too easily and too often overlooked; this series will go a long way to redressing the balance. Anyone with an interest in British history in general or the Royal Navy in particular will find this series an invaluable and enjoyable resource.

**Tim Benbow**
**Defence Studies Department,**
**King's College London at the**
**Defence Academy of the UK**

# Acknowledgements

To try and distil over five centuries of British imperial and naval history into one book has been a mammoth undertaking, and I could not have completed it without the intellectual and emotional support of a number of key people. Undoubtedly the biggest shout out has to go to the NMRN's Duncan Redford, who advocated for the inclusion of this volume following our discussions about the idea, and, as series editor, he has been a galvanising and sympathetic voice keeping my eye on the prize, whose positive and constructive feedback gave me a much-needed boost to reach the finish line. His patience and understanding has been shared by the publishers, I.B.Tauris, and Jo Godfrey in particular. Also at the NMRN, my thanks go to Stephen Courtney for his assistance in identifying suitable images from the museum's collection.

I am forever indebted to my boss, Ian Phimister, for his trust in giving me an opportunity in academia with the time and freedom to write this book, and whose loyalty and support I endeavour to repay. Other colleagues at the University of the Free State also deserve a huge amount of credit for tolerating many interruptions of their own important work to discuss my progress or problems and the petulance which often accompanied that, in particular Kate Law, Rosa Williams, Tinashe Nyamunda, Clement Masakure, David Patrick, Ilse Le Roux, Andrew Cohen (who cast his expert eye over parts of Chapter 7) and Lindie Koorts (for introducing me to South Africa's four-legged naval hero 'Just Nuisance'!). I also wish to express my gratitude to the Prestige Scholars Programme at UFS for inviting me to join their writing retreat in August

2014, where I completed most of Chapter 5 and had stimulating conversations with Neil Roos, Jackie du Toit and Bradley Smith especially, which helped refine and build confidence in my ideas.

The scale of this subject could almost merit its own series, and to attempt to do it justice in one volume has required standing on the shoulders of many giants, some of whom have been kind enough to spare me their time and expertise in emails and in person, notably Ashley Jackson, Clifford Pereira, Jonathan Rayner, Cindy McCreery and Steven Gray, and their pioneering work inspires my own contribution to the scholarship. I am most grateful to Stephen Locks for kindly allowing me to share the photographs of his grandfather, Able Seaman Albert Edwards Scott, who served with the Special Service Squadron.

Finally, there are many valued family and friends outside of history, but particularly my wonderful partner Ruth, parents Lynne, Phil and Bonnie, brothers Oliver and Cameron, and friends Jack, Chardé, Luke and Lee, whose interest in my personal and professional wellbeing, willingness to listen and unwavering belief in my ability continues to give me strength to tackle the considerable challenges that come with this career.

# Introduction

In the space of just over 300 years, a small island on the edge of Europe rose to command a quarter of the Earth's territory and a fifth of its people. As the largest empire in history, one on which 'the sun never set', its influence stretched even further. Yet to rule the world Britain had to first 'rule the waves', a task only made possible by the Royal Navy, producing a symbiotic relationship where the colonies materially strengthened the senior service so that it could strengthen the empire further. This is not primarily a story about the heroic deeds of 'great men' leading mighty fleets in battles for national survival or the British statesmen who rose and fell with them. Though colonies were among the spoils of victory, it was not enough to conquer a territory; to retain it a nation must also control its inhabitants. For that, internal campaigns waged by smaller vessels were much more significant for projecting an impression of Britain's omnipresence, which psychologically underlined its imperial power and contributed to the 'colonisation of the mind' and spread of British ideology. This manipulation of the cultural battlefield to win the psychological war continues to influence contemporary occupations by foreign forces, most recently in Afghanistan and Iraq in the early twenty-first century, where social scientists were embedded with the military to help map the 'human terrain' they operated in.[1]

In the century after Napoleon's defeat, when Britain's empire expanded the most under the almost ironically named *Pax Britannica* ('British Peace'), the Royal Navy was a key psychological weapon to sway sovereign states into signing deals weighted towards British commerce

1

and strategy. This tactic would become known as 'gunboat diplomacy', and it was built upon impressions of British prestige which allowed the Royal Navy to wield both the carrot and stick of intimidation and friendship. Mental valuations of prestige were vital to Britain's imperial and 'Great Power' status, and as Britain's most visible and persuasive global ambassadors, the Royal Navy played a front-line role in its cultivation. Imperial power was exerted through more than just gunfire, with the navy 'showing the flag' to win 'hearts and minds' through displays of 'naval theatre', wooing local communities with spectacles of technological and professional superiority which simultaneously impressed and dissuaded dissenters from offering cause to fire those guns in anger.[2]

The Royal Navy's stabilising presence created a conducive environment for missionaries and merchants to carry out their business of converting 'native' peoples to Britain's religious and economic way of thinking. Naval security facilitated globalisation and the international migration of goods, people and ideas. It altered the colonies' cultural map by inspiring new forms of artistic expression, popular entertainment, the adoption of organised sports and changes in dress, aspirations and behaviour. Naval officers provided models against which colonial communities measured their 'Britishness' and the social status which accompanied that. Writings and depictions of the navy's imperial exploits served a similar purpose for people in Britain, inspiring literature, paintings and theatre, contributing to a construction of the nation's identity which highlighted its cultural and scientific achievements compared to the more 'primitive' races the navy encountered during its voyages. This provided a moral justification for colonial rule and its professed paternalistic mission to improve the social condition of indigenous people – by teaching them 'British values' – which would find expression in the Royal Navy's training and development of colonial naval forces as they evolved into sovereign navies within the Commonwealth.

Beginning with the intellectual seeds that sparked England's early colonial ventures and the Royal Navy's growth, its scientific explorations and 'exotic' encounters with unfamiliar cultures that provided ideological justifications for imperialism, the naval diplomacy and campaigns that expanded Britain's empire and their cultural impact at home and abroad, to the development of colonial navies and the empire's naval response to international pressures, world war and decolonisation. This book explores how the Royal Navy was one of the most influential institutions not just in the history of Britain, but in the history of the modern world.

Fig. I.1. The British Empire (and remaining Overseas Territories underlined)

CHAPTER 1

# An Empire Emerges

Though the political entity of 'Great Britain' would not emerge until the 1707 Act of Union, the intellectual foundation of a 'British Empire', built upon a strong Royal Navy, developed over the preceding centuries. The royal courts of Tudor England brought together scholars, wealthy aristocrats, entrepreneurs and sailors who petitioned the monarchy to back colonial ventures in the 'New World' by providing a navy that could protect their interests against Spanish and Portuguese rivals. From these networks emerged chartered trading companies who relied upon the expertise of naval captains to lead their colonising expeditions. In the process, unknown lands were discovered and settled, creating new resources and markets for the English economy. State intervention to stave off foreign competition strengthened the navy by stimulating shipbuilding and training seamen, allowing it to seize additional colonies from rival nations in wartime. A symbiotic relationship developed between the Royal Navy and the empire, as trade needed naval protection, and warships required safe harbours to refit and resupply with colonial men and resources, leading to annexation for strategic as well as economic purposes. Yet imperial and naval power was psychological as well as physical, and indigenous encounters gave rise to racial ideologies which justified British authority over indigenous peoples, both within the colonies and the 'Senior Service'.

## The fifteenth and sixteenth centuries

By the end of the Hundred Years' War, England's territories on the European mainland had been reduced to just the Pale of Calais. During

1436, following the siege of that city and with the English having fled Paris, the first English language political poem emerged. While the country's aspirations for continental domination were dwindling, 'The Libelle of Englyshe Polycye', or 'Little Book of English Policy', was 'exhortynge alle Englande to kepe the see enviroun'. It prophesised that as an island nation, England's future prosperity would depend on it acquiring and maintaining overseas colonies, particularly Ireland, for which strength at sea would be crucial: 'wyth alle your myghte take hede / To kepe Yrelond, that it be not loste'. Though unappreciated at the time, 'The Libel' marked the beginning of a naval intellectual tradition within English political debate, inspiring the writings of prominent navalists over succeeding centuries, such as Richard Hakluyt and Samuel Pepys, and used as recently as 2012 to emphasise the ongoing importance of naval power to Britain's national and foreign policy.[1]

While Henry VII (having turned Columbus down) backed John Cabot's transatlantic voyages which gave England its historic claim to Newfoundland, it was during the reign of his son Henry VIII that the monarchy began to take a serious interest in the country's navy. In 1511, the year the *Mary Rose* was launched at Portsmouth, his councillors argued that England's maritime geography demonstrated that it was God's will that the country should expand across the Atlantic Ocean and establish colonies in the New World, as Spain had done to its profit:

> Let us in God's name leave off our attempts against the *terra firma*. The natural situation of islands seems not to consort with conquests of that kind. England alone is a just Empire. Or, when we would enlarge ourselves, let it be that way we can, to which it seems the eternal Providence hath destines us, which is by sea.[2]

In 1577, the first recorded reference to a Royal Navy and 'Brytish Impire' appeared in John Dee's *General and Rare Memorials Pertayning to the Perfect Arte of Navigation*. Dee was a notable mathematician, astronomer and occultist, and close advisor to Queen Elizabeth I. He was also a keen advocate for English imperial expansion, and used classical historians, medieval chroniclers, Renaissance cosmographers, Papal Bulls, genealogical charters and maps to press the Queen's claim to 'a great parte of the sea Coastes of Atlantis [otherwise called *America*] [ . . . ] and of all the Iles nere unto the same [ . . . ] and Chieflie all the *Ilands Septentrionall* [Greenland and Friesland]'. At the time that *General and Rare Memorials* was published, England found itself at the mercy of pirates, while foreign fishermen pilfered catches from her waters. In writing this book, Dee hoped to convince the Queen to create a 'NAVY-ROYALL' of 'Three score Tall Ships, (or more)' for the purpose not only of protecting the country from threats close to home, but to establish a British maritime empire:

> I come to my chiefe purpose [ . . . ] to stir upp your Majesties most noble hart, and to directe your Godlie conscience, to undertake this Brytish discover, and recovery Enterprise, in your owne Royall Interest: for the great good service of God, for your highness immortal fame, and the marvailous Wealth Publick of your Brytish Impire.[3]

This idea was graphically illustrated on the book's woodcut title page. Elizabeth is depicted directing the ship of state towards the Roman goddess Fortuna, who is holding out a laurel wreath for the Queen. The wreath was synonymous with the great empire of the Romans, worn by its Caesars to represent their supreme imperial authority, and it was now being passed to Elizabeth. During the Middle Ages, Fortuna was associated with both a ship's rudder, as she steered fate, and a cornucopia or horn of plenty, and thus her presence denotes that it is Elizabeth's destiny to found a maritime empire for the nation's prosperity. Britannia is seen kneeling on the shore and beseeching the Queen to build a 'fully-equipped expeditionary force' for that purpose. To her left is an ear of wheat, the promise of food abroad as the kingdom suffered from corn shortages, while also representing a Hermetic symbol for man. Its life-giving qualities are juxtaposed with the presence of a *memento mori* in the form of a skull to the right, where Elizabeth's ship emerges. This warns of the death of England if the Queen does not follow her European rivals' example by acquiring an overseas empire, the natural resources and the physical and spiritual nourishment that this offered the nation and its people. The presence of the archangel St Michael demonstrates that Elizabeth's mission is sanctioned by God, 'the HEAVENLY KING, for these many yeres last past, hath, by MANIFEST OCCASION' made her his Protestant defender 'of the most parte of Christendome'. She was thus

Fig. 1.1. Woodcut from Dee's General and Rare Memorials Pertayning to the Perfect Arte of Navigation

charged with bringing light to the peoples of the New World, darkened by their indigenous 'heathen' practices and the heretical Catholicism of Spain and Portugal. This is visualised by the sun on St Michael's shoulder, chasing away the dark night and moon. The nine stars in the sky might also symbolise the nine celestial spheres, as referred to in Dante's *Paradiso* which recounts the soul's ascension to heaven and the final sphere, 'empyrean', from whence the word 'empire' comes. Thus, in forging a British Empire with a Royal Navy, Elizabeth would get closer to God by creating a heaven on Earth.[4]

Dee's voice was not a solitary one at this time. The prominent London printer John Wolfe published a translation of the Dutch Protestant traveller Jan Huygen van Linschoten's *Discours of voyages into ye Easte and West Indies* in 1598, which he dedicated to the 'Judge of the High Court of the Admiralty' in the hope that 'you shall finde him any way beneficiall to our Countrey and Countrey men'. Its message was again an imperial one of religion and trade, promoting England's 'Honour over all '*Countreys* of the *World*' and advancing

> the Credite of the Realme' by employing 'our Wodden Walles [ . . . ] in forraine partes [ . . . ] for the dispersing and planting true Religion and Civill Conversation therein: as also for the benefite and commodity of this Land by exportation of such thinges wherein we doe abound, and importation of those *Necessities* whereof we stand in Neede.[5]

This intrinsic connection between the navy, commerce, empire and power was also emphasised by the famous naval commander Sir Walter Raleigh at the turn of the sixteenth century, when he wrote: 'Whosoever commands the sea commands the trade; whosoever commands the trade of the world commands the riches of the world, and consequently the world itself'. Raleigh was a patron of Richard Hakluyt, who penned several important works arguing for English colonisation, including his *Discourse of Western Planting*, which he presented in 1584 to Queen Elizabeth in support of Raleigh's scheme to establish an English colony in the New World. This marked a significant ideological break in that the Americas had previously been seen as a resource to be exploited and exhausted, in the privateering tradition, but Hakluyt conceptualised a much more modern form of settler colonialism, viewing it as a natural extension of England's borders and population overseas.[6]

Hakluyt dedicated to Raleigh his 1584 translation of René Goulaine de Laudonnière's *L'Histoire notable de la Floride*, in which he calls for England to follow 'the course that both the Spaniards and Portugals tooke in the beginnings of their discoveries and conquestes'. His most famous work, *The Principal Navigations, Voyages, Traffiques and Discoveries of the English Nation*, furthered the idea of a maritime empire by demonstrating a long-running tradition of English nautical endeavour, lending legitimacy to territorial claims by its sailors who 'in all former ages [ . . . ] have bene men full of activity, stirrers abroad, and searches of remote parts of the world'. Nor were colonial horizons confined to the Atlantic, for according to the courtier and poet Philip Sidney, Hakluyt 'hath served for a very good Trumpet' supporting the colonial projects of Humphrey Gilbert, including his expeditions to discover a Northwest Passage. Such a route would encourage 'commerce and traffike' between England and Asia, as argued for in Hakluyt's dedication to Francis Walsingham in *The Principal Navigations*' 1589 edition. This relationship was considered mutually beneficial, for while the English would 'use and exercise common trade with their marchants', they would simultaneously spread 'the incomparable treasure of the trueth of Christianity, and of the gospell' to Asia. An Oxford Theologian, Hakluyt stressed this religious imperative for English imperial expansion in his translation of Pietro Martire's *De Orbe Nouo decades*, again dedicated to Raleigh:

> There yet remain for you new lands, very ample kingdoms, unknown peoples [ . . . ] to be revealed, and by the good auspices of your arms and daring brought quickly and easily under the scepter of our serene highness Elizabeth, empress of the Ocean sea [ . . . ] God will be with you, seeing that the glory is intended for God himself, for the salvation of infinite souls, and for the increase of the Christian republic [ . . . ] nothing more glorious or deserving of honor can be passed to posterity than to tame the barbarians, to call back those who are in a natural state and pagan to the fellowship of civil life, to lead savage men back to within the orbit of reason, and to imbue atheists and others estranged from God with reverence for his divine will.[7]

Court intellectuals such as Dee, Wolfe and Hakluyt saw the creation of a maritime empire as more than an economic market for English imports and exports; it was also a divine and moral mission to 'civilise' and convert to Protestantism those less 'enlightened' parts of the globe. Such notions of a 'civilising mission' would continue to inspire British imperialists in the centuries that followed.

Several colonial ventures were attempted during this period, though they achieved little success. During his circumnavigation of the globe between 1577 and 1580, documented in *The Principal Navigations*, Francis Drake made landfall on the coast of northern California and claimed it for England as 'Nova Albion', but its geographic isolation made it impractical to settle. Humphrey Gilbert's attempts to establish a colony in Newfoundland in 1583 ended with his ship's sinking and his death. Of

Fig. 1.2. Drake's hat being stolen by 'Indians' near Rio de la Plata, c.1578

Raleigh's colonial ventures in Virginia, the first, promoted by Hakluyt, was abandoned after just a year, while the second was disrupted by the Spanish Armada and its colonists' unaccounted disappearance.

Both the navy and these early attempts at colonisation suffered from the absence of any sustained national maritime policy at this time. *General and Rare Memorials* reached some of the most influential privy councillors of the day, and many of Elizabeth's key advisors shared Dee's imperial vision theoretically. Yet his plans meant building the world's largest navy – comprising six fleets and establishing provisioning ports across the globe – and were deemed too expensive at £200,000 per year. The Queen herself was criticised by her admirals for not fully appreciating the opportunities that sea power presented and the strategic value of capturing overseas bases like the Azores. Yet, with an annual income amounting to just an eighth of that available to King Philip II of Spain, and no long-term revenues to borrow against, both the Royal Navy and colonial entrepreneurs were reliant on private investment to stay afloat. This,

Fig. 1.3. The 'Indians' of California greet Drake

and the privateering impulse of personal profit shared by sailors such as Raleigh and Essex and their financial backers, meant that their interests often diverged from those of the state, contributing to the failure of Drake's 1589 expedition to La Curuña.[8]

## The seventeenth century

It was during the early part of the seventeenth century that English colonisation of the New World really began to take root, with the organisation of private interests into chartered trading monopolies. A permanent settlement was finally founded by the Virginia Company in 1607 at Fort James (later Jamestown), with Hakluyt as one of its principal grantees. Whereas earlier efforts had failed largely because of their lack of logistical support, the Virginia Company utilised naval expertise to support its colonisation efforts. In 1609 it appointed Admiral Sir George

Somers of the Royal Navy to command its Third Supply Fleet, but en-route to Virginia a storm separated his flagship, the *Sea Venture*, from the rest of the fleet, wrecking it on the island that would become Bermuda. Having salvaged materials to construct two new vessels, the *Deliverance* and *Patience*, Somers and his crew arrived in Jamestown ten months later to find the colony close to starvation. They returned to Bermuda for food, but during the journey Somers became ill and died. His chance discovery of the island laid the seed for a colony there, and it was formally settled by the Company in 1612 as the Somers Isles. Around the same time, the Massachusetts Bay Company led colonisation efforts in New England, while the East India Company began to establish an English presence in the Indian subcontinent from 1611.[9]

Following the defeat and execution of Charles I in the English Civil War, a Commonwealth of England was declared on 19 May 1649, and the navy lost its royal association. With the country now a pariah state, English traders operated at greater risk in European waters, and pressed Parliament to support their interests. This and the threat posed by Prince Rupert of the Rhine, led to the expansion of the English Republican Navy, which played an important role in Oliver Cromwell's conquest of Ireland and the suppression of Scotland. It blockaded Rupert's flotilla in Kinsale (preventing it from landing the Royalist army in England) provided logistical support to Parliamentarian campaigners in Ireland, lent suppressive fire along the coast, cleared privateers and escorted merchantmen carrying supplies and reinforcements across the Irish Sea.[10]

The importance of maritime supremacy to national prosperity became enshrined in the principles of English mercantilism. The increase in successful colonial ventures provided a boom for international shipping, which needed to keep the colonies resupplied while transporting their commodities for export back to Europe. It was the Dutch monopoly in this business that allowed the Netherlands to rise to become England's major imperial rival in the sixteenth century. Profits generated by England's nascent colonies were being siphoned by duties imposed by foreign vessels. For English colonialism to be economically viable, it had to become self-reliant in shipping. It was this concern that led to the first Navigation Act being passed in 1651, arguably the most important piece of legislation for establishing the foundation of Britain's future hegemony.

It tied all overseas colonies to England by making them subject to Parliament, establishing a formal relationship of dependency and the political structure for a British Empire. The act also demanded that the right to trade with those colonies be retained solely for English shipping. This was used to justify the capture of Dutch vessels encountered near British territories, prompting the Netherlands to increase the armament of its merchantmen which escalated into the First Anglo-Dutch War of 1652–4. The English captured between 1,200 and 1,500 Dutch vessels during the conflict, but could not force a decisive victory or supplant the Netherlands as the world's major trading nation before the Peace Treaty of Westminster was signed on 5 April 1654.

It was not long, however, before England was at war again, this time allying itself with the French in their ongoing conflict with Spain. It was England's first 'imperial' war, with Cromwell's primary motivation being the acquisition of Spanish colonies in the Caribbean, particularly Hispaniola (Cuba). It was also the first war where England's colonial ambitions were furthered through military campaigning and Cromwell's so-called 'Western Design'. Whereas James I recognised the economic value of colonialism and supported the efforts of founding companies by granting trading monopolies, Cromwell was the first English head of state to appreciate the geopolitical and strategic value of depriving rival nations of their colonies through assertive action. Naval power was essential for pursuing this aggressive imperial policy. A fleet was dispatched from Portsmouth in December 1654, arriving in the Caribbean the following month. Although the attack on Hispaniola was a failure, in May 1655 the fleet did capture Jamaica, which would become Britain's largest and most valuable sugar-producing colony in the West Indies.

Meanwhile, General-at-Sea Robert Blake took action against the Spanish in the Mediterranean, capturing treasure fleets on 19 September 1656 at the battle of Cadiz and the battle of Santa Cruz de Tenerife on 20 April 1657. More significantly, Blake established a tradition whereby the navy furthered England's geopolitical ambitions not just through military engagement, but as a diplomatic tool to leverage the country's international status. Similar to the 'gunboat diplomacy' of the nineteenth century, which would spread Britain's 'imperialism of free trade', Blake used his fleet to extract treaties and secure naval base rights from the rulers of Tetouan and Tangier and gain compensation from the Duke of Tuscany and the Pope for supporting Royalist privateers, increasing the confidence of British traders who expanded their interests in the Mediterranean. A treaty was also signed with Portugal in 1654 which gave England trading concessions in Brazil, Bengal and West Africa in exchange for naval protection, while the country's elevated global status acquired by the fleet attracted calls for help and alliance from Morocco and Transylvania.

Regularly practiced two centuries later as part of the *Pax Britannica*, Blake's fleet first demonstrated that 'prestige' and the psychological impression of naval strength could be just as effective as the physical projection of that power in furthering England's political and economic ambitions. In seeking political influence over extra-European polities, an imperial policy beyond territorial annexation also emerged at this time. This notion of an 'informal empire' would again be pursued more vigorously during the 'new imperialism' of the nineteenth century.[11]

These wars demonstrated the importance of naval power and the strategic benefits that mercantile strength could bring to it, and additional legislature was passed which tightened the maritime bond between England and her colonies. The Navigation Act of 1660 ensured that three-quarters of all ships' crews were English, and colonial produce such as tobacco, cotton and sugar could only be transported to England and its territories. The 1663 Navigation Act meant that all European exports to the Americas had to pass through English ports first and pay duties on such commodities. Colonial trade thus provided a vital stimulus for English shipbuilding and the training of English seamen, which the navy drew upon in times of war.

Soon enough, a second war with the Netherlands broke out in March 1665. A crucial factor behind this was the influence that the navy and colonial traders enjoyed with those in power. The monarchy had been restored with King Charles II in 1660, who had appointed his brother, James Duke of York, as Lord High Admiral of the new Royal Navy. The Duke of York also headed the Royal African Company, and wanted to supplant the Dutch West Indies Company in the West African slave trade. In late 1663, Captain Robert Holmes of the Royal Navy was despatched by the King to seize Dutch trading posts in West Africa, culminating with the capture of Cape Coast Castle on the Gold Coast on 1 May 1664. Meanwhile, on 27 August 1964, four frigates under Captain Richard Nicholls sailed into New Amsterdam harbour, prompting the colony's peaceful surrender. Despite these early gains the Dutch rallied, sending a fleet under Michiel de Ruyter to retake all of Holmes' prizes apart from Cape Coast Castle, before raiding England's American colonies. By 1667, financial restrictions exacerbated by the plague and the Great Fire of London meant that the major part of the English fleet was laid up at Chatham. De Ruyter took advantage of this in June to launch his raid on the Medway, resulting in the destruction of three capital ships, the *Royal James*, *Royal Oak* and *Loyal London*, one First, three Second and three Third Rates, and the capture of the flagship *Royal Charles*.

After a month of blockading, the Treaty of Breda was signed, meaning England lost sugar-producing Surinam. In exchange, they were allowed to keep New Netherland (including New Amsterdam) – which became the colonies of New York (renamed after the Lord High Admiral), New Jersey, Pennsylvania and Delaware – as well as Cape Coast Castle, which gave them a foothold in the lucrative West African slave trade. Thus, though England came to the negotiating table defeated and humiliated, it gained a number of key possessions which became increasingly important for the empire's future growth and prosperity. In support of this, John Dryden published his *Essay of Dramatick Poesie* in 1668, in which he emphasised that national prestige, prosperity and empire depended upon naval supremacy:

> It was that memorable day in the first summer of the late war when our navy engaged the Dutch – a day wherein the two most mighty and best appointed fleets which any age had ever seen disputed the command of the greater half of the globe, the commerce of nations, and the riches of the universe.[12]

This followed 1667's *annus mirabilis*, where Dryden again justified the war as one 'most just and necessary' for exercising the nation's greatness by expanding its territorial possessions.

By the end of the seventeenth century, almost a third of England's imports arrived from the America and Asia, while the re-export trade of these goods to Europe, stimulated by the third Navigation Act, amounted to around 40 per cent of the country's total exports. Among these, tobacco imports boomed from 50,000 lb in 1615, to 38,000,000 lb in 1700. England became a fiscal-military state as custom revenues acquired through the Navigation Acts rose and were reinvested into the navy which enforced that legislation. It had a cumulative effect, as this naval guarantee encouraged more traders to speculate in shipping ventures, while increased demand from merchants and the Admiralty stimulated a boom in shipbuilding.[13]

In the literary tradition of Dee and Hakluyt, John Evelyn published his *Navigation and Commerce* in 1674. Originally intended as the preface for a history of the Anglo-Dutch Wars commissioned by King James II, Evelyn reflected the change in political attitudes that had taken place during the seventeenth century. Rather than leave maritime trade to private merchants, it was now appreciated that the State had to take a more active role in promoting and protecting England's commercial interests. Echoing Raleigh, Evelyn argued that the strength of the nation, its international reputation, and its economic and imperial power, were dependent upon possessing a powerful navy:

> To pretend to universal monarchy without fleets was long looked on as a politick chimaera [ . . . ] whoever commands the ocean, commands the trade of the world, and whoever commands the trade of the world, command the riches of the world, and whoever is master of that, commands the world itself.[14]

Colonies played an intrinsic role in this relationship, supplying raw materials, re-exports and markets for English produce, while providing the infrastructure required for protecting that trade in the form of safe harbours, provisions for the replenishment and repair of Royal Navy vessels, and manpower for its crews. As part of the State's bargain, by the 1680s two or three Royal Navy cruisers were regularly deployed in the Caribbean. From the plundering nation of privateers during the sixteenth century, colonisation and the Navigation Acts had given Britain a larger stake in legitimate trade, and the navy consequently became involved in anti-piracy measures to protect this. The growing economic importance of the Caribbean was further emphasised when the English government despatched Rear Admiral John Benbow to the region with a squadron of men of war during the 1689–97 war.[15]

As the colonies' wealth grew, so did their influence upon decision-making in Britain. With London the financial centre of this maritime imperial network, those commercial groups invested in colonial trade, plantations and shipping were able to lobby government and steer the navy's course to prioritise their interests over home defence. When deployed on colonial stations, Royal Navy Captains could be undermined by local governors who had a tendency to view their warships from a land-based perspective as stationary harbour batteries, without appreciating naval doctrine and the importance of intercepting an enemy on the high seas. The authority of colonial Governors in local naval matters was confirmed in 1697 by the Council of Trade and Plantations and by royal decree:

> The Commanders of his [Majesty's] ships that are sent for the service and defence of any Colony [are] to be under the direction of the Governor of each of those Colonies during their continuance there. Also, when the commanders have occasion for seamen they must apply to the Governor, to whom the sole power of impressing seamen is entrusted.[16]

By the end of Queen Anne's reign in 1714, the Admiralty had re-established ultimate control over Royal Navy vessels on overseas stations. Yet this tension between local interests and imperial strategy would continue to plague colonial–naval relations into the twentieth century, becoming a major factor behind the colonies' pursuit for naval autonomy by forming their own navies.

## The long eighteenth century

The Royal Navy became the navy of Great Britain following the 1707 Act of Union between England and Scotland, resulting partly from the financial calamity of Scotland's failed scheme to establish a colony on the Central American Isthmus of Darien. As the 1660 Navigation Act barred Scotland from trading with England's colonies, a 'Company of Scotland Trading to Africa and the Indies' was established and sent a fleet from Leith with five ships and 1,200 men to found the colony in July 1698. Yet upon their arrival the colonists soon succumbed to dysentery, fever and Spanish attacks, with only one vessel making it home. By March 1700, a second expedition had also surrendered to the Spanish, costing the cash-strapped country £200,000 and hundreds of lives in this futile colonial venture. Following the political union, the sailors and three vessels of the Royal Scots Navy were amalgamated with England's, though some Scottish officers instead joined Peter the Great's Russian Navy.[17]

By the late 1720s, the Royal Navy's anti-piracy patrols had allowed most transatlantic merchantmen to dispense with carrying deck guns for their protection, while peacetime insurance rates also fell. It was able to maintain a constant Caribbean presence due to the development of facilities in Jamaica and Antigua which allowed the navy to clean and refit its ships without having to return to Britain. From 1749, a naval base was developed on the North American coast at Halifax. For the eighteenth-century economist Malachy Postlethwayt, this development of colonial infrastructure demonstrated the dynamism of Britain's imperial power compared with Spain's stagnant empire:

> Spain, indeed, has greater countries and more subjects in America, than we have, and yet does not navigate in that trade a tenth part of the shipping that we do. By a lucky kind of poverty, our dominions there have no mines of gold or silver: we must be and ought to be, contented to deal in rum, sugar, rice, tobacco, horses, beef, corn, fish, lumber and other commodities that require great stowage; the perpetual carriages of these, employ above 100,000 tons of shipping.[18]

Whereas Spain acquired immediate wealth by mining its territories of precious metals, Britain developed its colonies to produce a more sustainable and expansive economic system. Postlethwayt consequently argued that 'the Almighty place the true riches of this earth [sugar, rice, tobacco, etc.] on the surface of it', again connecting the economic and religious impulses of empire. The need to transport greater volumes of lower-value cargo to make a profit also stimulated the expansion of the British fleet over its earlier rivals. The development of shipping marked the transition from the old parasitic empires of Spain and Portugal, to the more symbiotic imperialism pursued by Great Britain, the Netherlands and France. As Thomas Lediard observed in his 1735 *Naval History of England*:

> our trade is the Mother and Nurse of our Seamen; Our Seamen the Life of our Fleet; And our Fleet the Security and Protection of our Trade: and that both together are the WEALTH, STRENGTH, and GLORY of GREAT BRITAIN.[19]

The protection of colonial trade did mean that the main fleet was occasionally weakened at home. This happened in 1745, when Jamaica's lucrative sugar exports were threatened by a build-up of French and Spanish ships and reinforcements had to be diverted from the Royal Navy's Mediterranean squadron, undermining its blockade of France.

Even within a mercantilist system, Daniel Defoe argued as early as 1707 that Britain did not need to annex a territory outright, it only had to develop economic links in order to exert power and influence over it: 'We want not the dominion of more countries than we have; we sufficiently possess a nation when we have an open and free trade to it'. Defoe's words were prophetic, as a century and a half later this principle would underpin the development of an 'informal empire' under Britain's 'imperialism of free trade'. Even during the eighteenth century, not all territorial acquisitions were intended to be permanent. Strategic considerations during war meant that the Royal Navy would take foreign colonies as leverage for the negotiating table to reclaim losses suffered elsewhere. This gave it access to prime harbours as staging posts, while denying them to rival fleets and privateers. Captured foreign trade also increased Britain's own profits and undermined the enemy's economic ability to wage war. Mercantilism actually meant that British plantation owners in the Caribbean opposed the incorporation of new colonies which produced the same commodity within this protectionist system.[20]

The Seven Years' War, between 1756 and 1763, marked an escalation in imperial rivalry[i]. Arguably the first 'world war', battles were waged and won over colonial possessions across the American, African and Asian continents, a feat only possible because of the Royal Navy and the maritime imperial network cultivated over the forerunning century. As

---

[i]   See also Martin Robson, *A History of the Royal Navy: The Seven Years' War.*

the renowned American naval theorist Admiral Alfred Thayer Mahan later observed:

> The one nation that gained in this war was that which used the sea in peace to earn its wealth, and ruled it in war by the extent of its navy, by the number of its subjects who lived on the sea, and by its numerous bases of operations scattered over the globe.[21]

The conflict proved the value of possessing a system of overseas naval bases that allowed the Royal Navy to stay resupplied, repaired, and able to fight for longer than its French and Spanish counterparts. Even so, Europe remained at the centre of Britain's imperial power and naval strategy, as Admiral George Anson remarked in 1756:

> Our colonies are so numerous and so extensive that to keep a naval force at each equal to the united force of France would be impracticable with double our Navy. The best defence therefore for our Colonies as well as our coasts is to have such a squadron always to the Westward as may in all probability either keep the French in port or give them battle with advantage if they come out.[22]

This strategy culminated in the battle of Quiberon Bay on 20 November 1759, and, even before the defeat of the French fleet, allowed the Royal Navy to operate freely in North America. With the severance of its supply lines from Europe, New France fell to Britain. At the Peace of Paris, a victorious Britain kept the colonies it captured in Grenada, Dominica, St Vincent, Tobago and Senegal, while its pre-eminence in Canada and India was also confirmed by France's eviction. These latter territories would become vitally important to the 'Second British Empire' of the nineteenth and twentieth centuries.

Britain's interests and ambitions were now truly global, stretching beyond the transatlantic world which the First British Empire had centred on. This was emphasised in the distribution of its trade, and the growing importance of the East, particularly India and China, for their rare commodities and as markets for Britain's revolutionising manufacturing industry. With the loss of the 13 American colonies,[ii] this reorientation became even more important, as was Britain's ability to protect those

---

[ii] See also Martin Robson, *A History of the Royal Navy: The American Revolutionary War*.

Eastern commercial interests. With this in mind, Commodore George Johnstone was dispatched in 1781 with a squadron of two ships of the line, three Fourth Rates, and transports, to capture the Dutch colony at the Cape of Good Hope, the strategic hub of East Indian trade. Johnstone was thwarted by a French fleet led by Admiral Pierre André de Suffren, who had uncovered the British plan and attacked the unwitting Royal Navy squadron while it was anchored at Cape Verde. French reinforcements and naval superiority around the Cape caused Johnstone to call off the operation.

With the outbreak of the French Revolutionary Wars in 1793, the directors of the East India Company put pressure on the British government to secure the Cape route, calls which intensified after the Netherlands sided with France in 1795. A task force was deployed under Admiral George Elphinstone, the First Viscount Keith, and on 16 September that year the Cape was captured following a three-month campaign. It was briefly handed back to the Batavian Republic in 1803 before Britain occupied it for a second time in 1806. This was ratified by the 1815 Treaty of Paris, which also ceded the Gambia, Sierra Leone, Ascension Island, Mauritius, the Seychelles, Ceylon and Malacca to Britain, a string of pearls along the lucrative East Indian trade route, with the Cape the jewel at the centre, virtually equidistant from Gibraltar, the West Indies, the Falkland Islands, India, China and Australia. In addition, Britain's position in the Caribbean was strengthened with the acquisition of St Lucia, Tobago and Guiana, in the Mediterranean by Malta and the Ionian Islands, and in the North Sea with Heligoland. In sum, the successful wars waged against France between 1756 and 1815[iii] provided Britain with the vital links in a logistical chain that allowed the Royal Navy to stay afloat and safeguard British trade worldwide, and facilitated the growth of its economic and imperial power during the coming age of *Pax Britannica*.[23]

## Colonial marines

Concurrent to the Napoleonic campaign, war broke out in 1812 with the United States, though not for two years would major forces arrive under

---

[iii]  See also Andrew Baines, *A History of the Royal Navy: The Age of Sail.*

Vice Admiral Sir Alexander Cochrane. Here Britain's abolition of the slave trade in 1807 provided the Royal Navy with opportunities to strengthen its manpower. Under humanitarian guise, though strategically motivated, Cochrane issued a proclamation on 4 April inviting American slaves to emigrate aboard British vessels, where they would be given the choice of 'entering into his Majesty's sea or land forces, or of being sent as FREE settlers to the British possessions in North America or the West Indies'. While wartime labour shortages meant former slaves had served with Royal Navy vessels on the North American station throughout the eighteenth century, they never exceeded 6 per cent of its crews, though some black sailors did occasionally rise to become officers. The most notable of these was Lieutenant John Perkins, who was given command of HMS *Endeavour* in 1782.[24]

In May 1814, Admiral Sir George Cockburn began recruiting a 'Corps of Colonial Marines from the People of Colour who escape to us from the Enemy's shore'. On 4 August 1814, the American Adjutant General was alerted to the effectiveness of the Royal Navy's recruitment efforts and the successes its Colonial Marines were achieving in ambushing the United States Army:

> Our Negroes are flocking to the enemy from all quarters, which they convert into troops, vindictive and rapacious – with a minute knowledge of every byepath. They leave us as spies upon our posts and our strength, and they return upon us as guides and soldiers and incendiaries.[25]

By the final week of May, over 200 former slaves assaulted an American battery at Pungoteaque, Virginia. A few weeks later they were used in raiding parties along the Patuxent river, while in August 300 contributed to the British victory at the Battle of Bladensburg, which resulted in the burning of Washington, DC.

The performances of the Colonial Marines drew praise from Admiral Cockburn, who believed that their African blood made them biologically suited to fight in hot and humid environments. This was significant in that it not only marked the first attempt by the Royal Navy to recruit and organise colonial manpower into a distinct homogenous unit, but also because it fostered the development of racial thinking around naval management. Imperial ideology, particularly the notion that some ethnic groups were naturally more 'martial' or suited to serve in 'tropical' environments, would gain credence over the coming century, as social Darwinism and scientific racism justified British domination over the indigenous peoples they colonised.[26]

On 13 August 1814, the Royal Marine Lieutenant Colonel Edward Nicholls, known as 'Fighting Nicholls', was assigned to train 500 Black and Native Americans for the Colonial Marines, about whom he expressed paternalistic opinions:

> Towards the Indians you will show the most exact discipline, you will be an example to those sons of nature, you will have to drill and instruct them, in doing so which you must be patient with and watch their likes and dislikes and be careful to offend them in nothing. Above all things sobriety must be your constant care, one example of drunkenness may ruin this. When the men of colour who are expected to join us arrive you will be strictly careful in your language and manners to them if they do not take your instructions as readily as you wish [ . . . ] remember they have been oppressed by cruel taskmasters and under slavery man's best faculties are kept dormant, what a glorious prospect for British soldiers to set them free.[27]

Cases like Lieutenant Perkins have been used to argue that the Royal Navy's attitude towards black seamen was 'liberal by the standards of the societies from which they had come', creating 'a world in which a man's professional skill mattered more than his colour'. While Nicholls encouraged his officers to show respect towards their non-white charges, his words nonetheless conveyed an air of racial superiority which defined nineteenth-century paternalism and justified Britain's imperial leadership over less-'developed' peoples. There is the racial stereotype of the Native American corrupted by alcohol, for whom the British had to set a moral example, while as 'sons of nature', they epitomised the 'noble savage' – 'hard primitivists' who were considered tough but uneducated. It was left to British officers to not only train, but 'civilise' these colonial men, using Christianity as the means to achieve their social and cultural elevation. In espousing common principles of honour, duty and loyalty, religion also instilled discipline by discouraging acts of retribution:

Show yourselves to be Christians by your deeds – mercy will cause the British men to love you, it will be a chief motive for their acknowledging you as brothers, write deep these words on the Tablets of your memory and look at them with serious and charitable resolution, when we in the possessions of your former taskmasters do them no other harm or violence than is necessary.[28]

In exchange for their service, the Colonial Marines were promised 'the comforts of enjoying rational liberty, solid property with the rights of a British Man', and after the war over 2,000 freed slaves settled in Trinidad and Nova Scotia, contributing to the growth of those colonies. These Enlightenment principles which portrayed Britain's empire as a liberal and just one, particularly compared to its European rivals, would increasingly be used to justify British imperial expansion over the coming century.[29]

CHAPTER 2

# Science and Exploration

The European Enlightenment was an age that sought to improve human knowledge, morality and society through reason, scientific method, and empirical investigation. Voyages of exploration brought new lands and people into contrast, against whom Europeans measured their own civilisation and sense of progress. Racial stereotypes of the 'primitive', 'savage' and 'native' emerged, as did an imperial mission to 'civilise' and raise the material and spiritual condition of those groups. Due to their relative isolation, the peoples of the Pacific were seen to epitomise this child-like innocence upon which notions of the 'noble savage' were built. They inspired paternalistic ideas which prescribed that Europeans should carry out the role of teacher and guardian, justifying imperialism as a philanthropic effort to ensure the progress and salvation of the colonised. Through its exploratory missions, the Royal Navy played a pivotal role in the construction of such 'knowledge', establishing contact, securing treaties and providing the lens through which those in Britain viewed other parts of the world via the logs, letters, journals, sketches and maps created by naval officers and their scientific companions.

Though Francis Drake was the first Englishman to sail the Pacific Ocean during his circumnavigation, it would take almost a century to pass before another compatriot followed him into this part of the world. In 1669, Lieutenant John Narborough was ordered by James, Duke of York, to take the *Sweepstakes* through the Straits of Magellan and into the Eastern Pacific, 'to make a Discovery both of the Seas and Coasts of that part of the World, and if possible to lay the foundations of Trade there', but after two years he achieved little success in this Spanish-dominated region. The next Royal Navy expedition to the Pacific was led by the renowned pirate William Dampier. Having corresponded with the First Lord of the Admiralty, dined with Pepys and consulted with Council of Trade and Plantations, he convinced the Admiralty to support an attempt to discover the hypothetical great southern continent of Terra Australis, on the assumption that it would be 'reasonable to conceive that so great a part of the World is not without very valuable commodities'. The Royal Navy gave Dampier a commission and provided him with a ship, the *Roebuck*. He made it to the coast of New Holland (Western Australia) and discovered the island of New Britain, before the deteriorating condition of both his ship and crew forced him to return home before the unknown east coast of Australia could be explored.[1]

Accounts of Dampier's voyages were popularly read in Britain and emphasised the exoticness of his encounters, describing the body and face painting exhibited by New Holland's Aboriginal people as

> adding very much to his natural deformity; for they all of them have the most unpleasant looks and the worst features of any people that ever I saw, though I have seen great variety of savages [ . . . ] blinking creatures [ . . . ] with the same black skins, and hair frazzled, tall and thin.

Seen as lacking houses, clothing, tools and religion – measures of European 'civilisation' – Dampier noted that 'setting aside their Humane Shape, they differ but little from Brutes'. By bringing into question their humanity, the indigenous peoples could be lowered to the level of a commodity, another colonial resource to be exploited, with Dampier believing that New Britain's 'strong, well-limb'd *Negroes* may be easily brought to Commerce'.[2]

For the first half of the eighteenth century, ongoing tensions with Spain meant that the Admiralty were reluctant to divert resources towards oceanic exploration. Ironically, it took the outbreak of war before two ships were sent to search for a passage from Hudson's Bay to the Pacific Ocean in 1741. Sir Charles Wager, the First Lord of the Admiralty, was swayed by the strategic advantages of such a discovery, which would allow the Royal Navy to launch surprise attacks on the unsuspecting Spanish in the Pacific. Christopher Middleton, who knew the region after having served in the Hudson's Bay Company, was commissioned to command the expedition. Meanwhile, Commodore George Anson was leading a squadron of eight ships and 1,854 men around Cape Horn on a mission to capture Callao, Lima, Panama, the Manila-Acapulco treasure galleon, and incite a Peruvian revolt against Spanish rule. Though Anson only succeeded in one objective, returning in 1744 having lost all but his flagship *Centurion* and 188 men, his circumnavigation and capture of the galleon seized the public's imagination and increased domestic interest in the South Seas as he advocated further explorations for the 'important purposes of navigation, commerce, and national interest'.[3]

Commodore George Byron, who had participated in Anson's voyage as a midshipman aboard HMS *Wager*, embarked upon his own circumnavigation with HMS *Dolphin* in June 1764. The Admiralty emphasised to Byron, as it did in other captains' instructions, that

> nothing can redound more to the honour of this Nation as a Maritime Power, to the dignity of the Crown of Great Britain, and to the advancement of the Trade and Navigation thereof, than to make Discoveries of Countries hitherto unknown.

This he pursued in 'discovering' the Gilbert Islands, Tuamotus and Tokelau, while also claiming the Falkland Islands for Britain on the basis that it was first discovered by John Strong of HMS *Welfare* in January 1690. Three months after its return, the *Dolphin* was once more dispatched to the Pacific on 22 August 1766, this time under the command of Captain Samuel Wallis, and was joined by the sloop *Swallow*, captained by Byron's former Lieutenant Philip Carteret. Shortly after leaving the Straits of Magellan, the two ships were separated; Wallis became the first European to establish contact with Tahiti in June 1767, while Carteret uncovered Pitcairn, the Duke of York Islands and the island group (in modern Papua New Guinea) that subsequently bore his name. In addition, Carteret's improved charting of the Makassar Strait and St George's Channel allowed for the establishment of greater trade with China. The information passed on by Wallis upon his return in May 1768 formed an important basis for Lieutenant James Cook's famous first

Fig. 2.1. Captain Wallis meeting Queen Oberea of Tahiti

voyage aboard His Majesty's Bark *Endeavour*, which departed on 26 August that year.[4]

Officially, Cook's mission was to record the rare transit of Venus across the Sun on 3 June 1769 for the Royal Society. Wallis' description of Tahiti, providing its longitude and latitude, presented the island as an ideal location from which to make the observation. This masked a secret set of orders given to Cook by the Admiralty, which stated that once Venus' transit had passed he was not to be 'diverted from the Object which you are always to have in View, the Discovery of the Southern Continent', which Wallis had also reported sighting. Imperial rivalry in the field of scientific and maritime discovery was one reason for the great secrecy attached to Cook's mission. British naval developments during the 1760s were closely monitored by an intelligence cell within the French Embassy in London, reporting directly to France's Secretary of State for War and the Navy, while Royal Navy movements were regularly printed in the *Gazette de France*. While sailing back to Britain, Carteret was passed by the *La Boudeuse* carrying the French explorer Louis-Antoine de Bougainville on his own circumnavigation. Exploration was not just about discovering 'Countries hitherto unknown' for humanistic purposes of scientific progress. It also exposed the naval condition of the nation's rivals by gathering intelligence on the disposition of strategic resources, harbours and fortifications. In this manner, Carteret was the first to reveal that Spain had established a fortified settlement on Juan Fernandez Island off the Chilean coast. Later in the century, reports that Jean François de Galaup, Comte de Lapérouse, was transporting 60 convicts to New Zealand to establish a French shipbuilding colony, expedited Britain's colonisation of Australia and led to the founding of the penal settlement of Botany Bay in 1788.[5]

Indigenous peoples were dispossessed by Admiralty instructions for captains like Cook to take possession of 'uncultivated' *terra nullius* ('land belonging to no one'), 'by setting up Proper Marks and Inscriptions, as first discoverers and possessors'. This meant depositing items such as coins, nails, axes, messages in bottles, carved inscriptions and flags. It was not insignificant that these markers should represent a level of social and scientific progress, in manufacturing and literacy, not witnessed among the indigenous populations, reinforcing European justifications for dominion based upon supposedly superior civilisation. Scientific advancement in the fields of cartography and astronomy provided Europeans with the tools to present the world in a manner which visually validated their imperial claims through maps, charts and tables drawn up by naval officers. Cook's maps of America's north-west coast, for instance, displayed few signs of indigenous life, and his officers described it as 'a space awaiting commercial development' by Britain. As has been graphically described, naval vessels represented the 'the inked needle of an instrument of enormous proportions that scribbles the shape of [countries] [ . . . ] in London', and drew the longitudinal links between the imperial centre and its colonial 'peripheries'. In this manner, maps played a role similar to industrial patents, as proof of original discovery and ownership.[6]

When 'natives' were encountered in new lands, Royal Navy officers were instructed to observe their 'Genius, Temper, Disposition and Number' to 'cultivate a Friendship and Alliance with them'. Though this was encouraged by coercion and 'making them presents of such Trifles as they may Value', in many cases it was secured through intimidation and demonstrations of British military superiority and firepower. It represented the sort of 'carrot and stick' tactic typical of 'gunboat diplomacy', which would gain prominence over the next century. In one instance, on 30 November 1769, when Maori claimed a buoy which was not for trading, Cook fired one of *Endeavour*'s guns over their heads, which made them 'meek as lambs' in subsequent negotiations.[7]

Royal Navy captains were regularly accompanied by botanists, such as Joseph Banks aboard the *Endeavour*, who were there to help carry out Admiralty instructions, to 'observe the Nature of the Soil, and the Products thereof; the Beasts and Fowls that inhabit or frequent it, the Fishes that are to be found in the Rivers or upon the Coast and in what Plenty'. Again, such investigations were not purely for the advancement of science; they gauged the material viability that new lands might have for British colonisation. Banks was a strong advocate for the penal colony in New South Wales and his findings were crucial to its establishment. This gave Britain a foothold for colonising the rest of the continent and its neighbouring islands of Tasmania and New Zealand, the latter having been first mapped by Cook. Although Cook's 1768–71 voyage would be the final one Banks accompanied personally, as President of the Royal Society with connections in government and the royal family, Banks became a central node in a network linking science, politics and the navy, which promoted exploration as a means of achieving imperial greatness. He ensured that most Admiralty expeditions also carried a scientific mission, with naval officers paying close attention to geomagnetism and oceanography, while collecting zoological, botanical and mineralogical specimens. Not all naturalists harboured imperial ideals; voyages provided them with an opportunity to acquire large collections of rarities which they could use for social advancement by forging reputations as 'gentlemen scholars' back in Britain. This followed a tradition since Dampier, who was elevated from the level of common pirate into the circles of respectable society by his observations of Pacific currents, meteorology and vegetation, presented through well-received books and the Royal Society.[8]

Dampier's *A New Voyage Round the World* (1697) had created a popular craving for Pacific travel literature and its tales of the 'exotic', that did not abate during the eighteenth century and which continued to influence British perceptions of national and racial identity. The voyages of Byron,

Wallis, Carteret and Cook were published together by Dr John Hawkesworth in a three-volume edition entitled *An Account of the Voyages undertaken by the Order of his Present Majesty for Making Discoveries in the Southern Hemisphere* (1773). It was a sensation, with four official and two pirated editions released in just two years, and the genre would inspire satirical works by Daniel Defoe and Jonathan Swift. Hawkesworth distinguishes these voyages from earlier ones by highlighting that they had not been made 'with a view to the acquisition of treasure, or the extent of dominion, but the improvement of commerce and the increase and diffusion of knowledge'. Mirroring this societal development, the word 'civilisation' entered common usage during the 1770s and is exemplified by Hawkesworth's Royal Navy representatives presented in stark contrast to the 'primitive' Pacific societies they encountered.[9]

Cook's voyages revealed an abundance of breadfruit on Tahiti and the other Society Islands, attracting interest from West Indian plantation owners who saw it as a potentially cheap and abundant food source for their slaves. They petitioned the government in 1787 to introduce the fruit to the Caribbean, securing the backing of King George III and his friend Joseph Banks, now President of the Royal Society, who recommended that Lieutenant William Bligh command the mission, having served on Cook's second Pacific voyage of 1772–5. This was an example of the fiscal-maritime state in action, with merchants soliciting networks of power to deploy the Royal Navy and advance their commercial interests. Imperial rivalry was crucial in gaining government and royal approval for the mission, as France was already in the process of transporting the breadfruit plant from Ile de France to its plantations in Saint-Domingue. The collier *Bethia* was purchased for the voyage, refitted and commissioned as HMS *Bounty*, which became infamous on 28 April 1789 for the mutiny of Master Fletcher Christian and others attracted to an idyllic life on Tahiti. After setting Bligh adrift in a boat with those still loyal to him, Christian led some of the mutineers and their Tahitian companions to Pitcairn Island, where they settled and burned the *Bounty* to avoid detection by the Royal Navy. After 47 days, and having navigated 3,618 nautical miles in the 23-foot open launch, Bligh made it to Timor and was returned to Britain. Having been honourably acquitted by court martial, he was assigned to command HMS *Providence* in August 1791 on a second attempt to complete the breadfruit mission. This time he successfully transported live specimens to St Helena, St Vincent and Jamaica, though the slaves refused to eat the food. Aside from this, the mutiny of the *Bounty* became a focal point for radical critics of British colonialism after Cook's voyages had raised sympathy for indigenous peoples and caused many to question violent annexation. Romantic authors became inspired by the tale of love on 'utopian' Tahiti, contrasting the freedom of the Pacific with the oppression and exploitation of Britain's colonies in the West Indies and India.[10]

A young midshipman serving under Bligh aboard the *Providence* was Matthew Flinders, who undertook two further voyages to the Pacific before becoming the first person to circumnavigate Australia during his third, as captain of the *Investigator* between 1801 and 1803. Though ostensibly a scientific mission to chart the coastline of New Holland, the *Investigator* was expressly dispatched to establish a British presence in Western and Southern Australia before the French, with whom Britain was again at war. Flinders popularised the name 'Australia', inspired by Alexander Dalrymple's *An Historical Collection of Voyages and Discoveries in the South Pacific Ocean* (1771), which had speculated upon the existence of an undiscovered Southern continent. Dalrymple had worked for the British East India Company before being made a member of the Royal Society in 1771, and though Cook was chosen ahead of him for the 1768 expedition, Dalrymple became the first hydrographer of the Navy in 1795.[11]

A resurgence in the Royal Navy's scientific mission followed peace with France in 1815. In 1825, Captain Phillip Parker King was instructed to take HMS *Adventure* and *Beagle* on an expedition to survey the South American coast and islands, but again there lurked an ulterior motive. The United Provinces of the Rio de la Plata (Argentina) and Chile had gained their independence from Spain, and the decline of that empire created a vacuum for British imperialism to exploit. King was tasked with establishing the commercial and strategic potential of these new countries, creating charts and sailing instructions for British merchantmen to open up trade with them. Imperial interests still accompanied the *Beagle*'s second voyage of 1831–6, made famous by the young naturalist Charles Darwin, but during the course of which Captain FitzRoy extracted a fine from Tahiti's queen that reinforced British hegemony over the island, while Britain's claim to the Falkland Islands was also reasserted.[12]

## The Northwest Passage

Cook's third and final voyage of 1776–9 had reignited the hunt for a Northwest Passage, begun by John Cabot in 1497 and continued by Martin Frobisher in 1576 and Sir Humphrey Gilbert in 1583 from which England's claim to Newfoundland was established. Accompanying Cook was George Vancouver, whose 1791–5 survey enabled the British settlement of British Columbia and Vancouver Island, though he too failed to find the passage. It was no coincidence that exploration targeted regions of commercial and strategic value that rival nations might exploit, and in this regard the Arctic offered a more economical route to Asia that was of particular interest to both the US and Russia. An active advocate of Arctic exploration in the first half of the nineteenth century who epitomised the network between imperial, naval and scientific interests was Sir John Barrow, Second Secretary to the Admiralty between 1805 and 1845. Barrow had previously worked for colonial diplomat Lord George Macartney in China and the Cape, and had collected commercial and strategic intelligence about the eastern seas and southern Africa for Henry Dundas, the former First Lord of the Admiralty, then president of the Board of Control and Secretary of State for War.

A member of the Royal Society and co-founder of the Royal Geographical Society in 1830 with the Royal Navy's Francis Beaufort and William Henry Smyth, Barrow was 'an ardent imperialist [ . . . ] convinced that Britain's security and future wealth depended on control of the world's sea lanes for trade and defence'. This belief saw him instigate a series of expeditions to trace the source of the Niger river, as well as a sustained search for the Northwest Passage after Banks received news of the ice cap's melting in 1817. A government prize of £20,000 provided an added material incentive for many distinguished Royal Navy officers to undertake the enterprise, including Sir George Back, Frederick William Beechey, Edward Joseph Bird, Edward Parry and Sir James Clark Ross. When fellow polar explorer Rear Admiral Sir John Franklin disappeared with HMS *Erebus* and HMS *Terror* in 1845, these men formed

Fig. 2.2. Stephen Pearce's *The Arctic Council Discussing a Plan of Search for Sir John Franklin* (1851)

part of the Arctic Council to pool their collective experience in the search for him. The Council was popularised by Stephen Pearce's 1851 painting, commissioned by his friend and keeper of the Admiralty records, John Barrow junior, which was exhibited in galleries throughout Britain.[13]

As with the Pacific voyages, the navy's Arctic expeditions captured the public's imagination and inspired romantic depictions of the ice, the polar sea, the Northwest Passage and the 'Eskimo', through art, poetry and popular literature such as *Moby Dick* and *Frankenstein*. Born in Halifax, Nova Scotia, Edward Belcher became assistant surveyor aboard HMS *Blossom* on Beechey's 1825–7 expedition to the Bering Straits, where he collected cultural artefacts from Eskimo communities. Like the Polynesians, the Innuit and Eskimo became seen as 'noble savages', having been geographically sheltered from the extravagancies of European civilisation. By comparing the material culture of these exotic 'Others' with that of prehistoric European civilisations, followers of Darwin's theories were able to place the Arctic communities within a progressive ladder of human evolution which provided a racist scientific rationale for British moral and intellectual leadership. Imperialism was justified on the grounds that indigenous peoples needed exposure to 'civilised' values in education, religion and morality in order to redress their socio-cultural lag.[14]

Paternalistic ideas about race and empire coincided with an evangelical revival which, following its suppression of the transatlantic slave trade, encouraged a belief that Britain and the Royal Navy had been tasked with a providentially-ordained moral mission to help spread the tenets of Christian 'civilisation'. Britain's geographic location as an island on the edge of Europe was seen as evidence of this, having compelled her people to venture forth onto the waves and achieve maritime and commercial supremacy, and they now had a moral obligation to share their good fortune, as the Society for the Diffusion of Useful Knowledge declared:

> The extension of civilization throughout every region of the earth where the people are ignorant and wretched, appears to be the peculiar duty which Providence has imposed upon a maritime and commercial nation.[15]

The *Journal of Civilization* proclaimed that the Royal Navy 'did not forget that they were also enlisted under the more glorious banner of the Lord of Hosts', and its officers were imbued with this sense of sacred duty through orders to 'report especially on the present state of those places and the Progress of the Inhabitants in Civilization and Christianity'. Many captains went further, actively helping missionaries by transporting them to their stations, donating funds and providing diplomatic and military assistance. HMS *Seringapatam*'s Captain William Waldegrave was described by Tonga's Methodist missionary George Turner as 'the decided friend of Missionaries [...] resolved to remove every impediment out of our way that lays in his power'. Commander Walter Croker of HMS *Favourite* offered military salvation to Tongatapu's Christian communities who were being attacked by pagan rivals; he informed the missionaries that 'the Providence of Almighty God had directed him in coming to the place', but was killed upon charging the fortifications. Influenced by Victorian morality, romanticism, neo-chivalry, and the cult of Nelson, many naval officers felt justified in exceeding their official orders to pursue a higher imperial calling. HMS *Daphne*'s Captain Edward Fanshawe vowed to be 'a sort of knight errant' during his Pacific tour, while Captain Francis Crozier, who died during

Franklin's final mission to seek the Northwest Passage, informed the Admiralty in 1836 that 'above all things, I hope I have exalted the character of the British Nation among uninstructed savages, by examples of Morality, firmness, and Moderation'.[16]

The navy's expeditions not only served to reinforce British cultural superiority over colonised peoples; being the first Europeans to venture into uncharted territories, both literally and figuratively, was also a way of demonstrating Britain's national prestige and technological superiority over its imperial rivals. Thus the British government funded a £150,000 expedition to discover the North Pole in 1875, led by George Nares in HMS *Alert* and *Discovery*, though it failed due to an outbreak of scurvy and inadequate clothing and equipment. Experiments with the submarine telegraph also attracted Admiralty interest, and HMS *Agamemnon* successfully laid the first cable across the Atlantic along with the USS *Niagara* in July and August 1858, after HMS *Cyclops* had surveyed the route between Ireland and Newfoundland. Though it only transmitted until September, a lasting connection was established in 1866, connecting Britain with North America a year later, and with India in 1870, Australia in 1872, and the rest of the world shortly thereafter, providing huge strategic, economic and political benefits for the Royal Navy, British trade and imperial unity. Britain's supremacy at sea and its monopoly over the submarine telegraph allowed it to respond more quickly to crises, whilst intercepting or severing enemy communications. When war broke out with the Zulus in 1879, a cable was laid between Durban and Aden to allow more central control over British troop movements, while in the first two months of the 1899 Boer War, 3,000 miles of cable were laid between the Cape of Good Hope and Cape Verde.[17]

## Coal-onialism

The Admiralty realised the practical advantages it could derive from scientific surveys, particularly in the field of geology after the advent of steam-powered vessels. It encouraged its officers to focus their observations in this direction by publishing *A Manual of Scientific Enquiry* in 1849, featuring contributions from Darwin and noted geologist Sir Henry Thomas De La Beche. Officers also received instruction at the

Fig. 2.3. A whale crosses the line as HMS *Agamemnon* lays the Atlantic telegraph cable in 1858

Museum of Practical Geology, and the School of Mines, where it was stressed that 'the discovery of mineral wealth is the most powerful incentive to the exploration and settlement of distant lands'. De La Beche was involved in the Admiralty Coal Enquiry between 1845 and 1850, evaluating potential fuel sources for the Royal Navy's conversion to steam power. With the Museum of Practical Geology, he tested the properties of British coal against specimens gathered by naval officers in strategically-located colonies such as Trinidad, Chile, Vancouver Island, Tasmania, Borneo and Bengal. Coal deposits at Nanaimo on Vancouver Island prompted Captain George William Courtenay of HMS *Constance* to wield his naval authority and declare British jurisdiction over that part of the island which now bears his name. Geologist Sir Roderick Impey Murchison, President of the Royal Geographical Society between 1843 and 1871, also used his scientific authority to further an imperial mission. An advisor to the Admiralty on charting mineral deposits, Murchison commissioned geological and topographical surveys to encourage colonial emigration, develop trade routes, strengthen naval defences and construct railways, while he was not averse to using gunboats to extend the empire by force when commercial penetration faltered.[18]

Royal Navy officers used the local knowledge they acquired while serving on colonial stations to personally profit from the empire's craving for coal. Such a collective invested in and developed the Nanaimo mines, which became an important fuel source for the navy until its conversion to oil in the early twentieth century. The Commanding Officer of the Pacific Station and HMS *Forward*, Lieutenant Commander the Honourable Horace Douglas Lascelles, created the Hareword Coal Mining Company in 1864. This was managed by Robert Dunsmuir, who later formed another coal-mining venture with the aid of over $5,000 from HMS *Grappler*'s Lieutenant Wadham Neston Diggle, $12,000 from Esquimalt's commander, Admiral Arthur Farquhar, and another $12,000 from Captain Frederick Wilbraham Egerton.[19]

With the exception of Nanaimo, the majority of colonial coal samples proved inadequate for the navy's use, they either produced too much ash and smoke, increasing warship visibility, or burned too quickly at temperatures too low for attaining high speeds. Most major Royal Navy stations continued to rely on exports of Welsh coal, either entirely or mixed together with local sources. When the China Station cut costs by switching to Australian coal in 1873, complaints from local officers saw the policy abandoned within three years and Welsh imports resumed. Only in 1882, with the discovery of Westport coal in New Zealand, would the Royal Navy find a colonial supply equal to that of Wales. Beginning with the Australia and China Stations, by 1900 New Zealand coal was powering Royal Navy vessels in the Pacific, Indian Ocean, East Africa, Cape of Good Hope and Americas. Naval demand also created improvements in colonial infrastructure, with several hundred thousand pounds invested in developing Westport's harbour and expanding storage facilities in Sydney.[20]

## Antarctica

Steam power opened up the world like never before, dramatically shortening trans-oceanic voyages and their susceptibility to the natural elements, while small river steamers penetrated areas otherwise cut off by difficult terrain such as dense jungle. Yet at the turn of the twentieth century, Antarctica remained an untameable frontier, barely explored

Fig. 2.4. HMS *Powerful* being coaled at Hong Kong, 1895

since James Ross' voyages with HMS *Terror* and *Erebus* in 1839–43. The Royal Navy's Commander Robert Falcon Scott was charged with leading an expedition there to recapture the public's imagination and distract them from the weariness of the Boer War. Jointly organised by the Royal Society and Royal Geographical Society, the expedition would also launch the Antarctic career of its third officer, Ernest Shackleton. It was hoped that the country might regain the prestige it had lost in that drawn-out colonial conflict by demonstrating how 'even in the last throes of an exhausting struggle, we can yet spare the energy and the men to add to the triumphs we have already won in the peaceful but heroic field of exploration' (*Morning Post*, 5 August 1901). Scott's vessel, the RSS *Discovery*, was specially constructed for the voyage, and, at its launch in Dundee on 21 March 1901, was celebrated as a 'magnificent ship [ . . . ] an embodiment of national enterprise [ . . . ] The purpose for which the ship is intended appealed strongly alike to the imagination and patriotic feeling' (*Dundee Advertiser*, 22 March 1901). Major companies like Cadbury's, Bird's and Colman's keenly donated their products to the expedition and advertised Scott's endorsements, including the benefits of

Fig. 2.5. Scott's party at the South Pole, January 1911

cooking seal meat in Bovril. Excitement about the expedition spread to
the far corners of the empire, with New Zealand offering 40 live sheep.
These were collected after *Discovery* left Britain on 6 August, symbolically
showing the flag at Cape Town and New Zealand, before reaching
McMurdo Sound in Antarctica on 8 February 1902 (named after
Lieutenant Archibald McMurdo of HMS *Terror*).

When Scott returned to Antarctica in 1910 on his fateful *Terra Nova*
Expedition, he declared that his mission was to be the first 'to reach the
South Pole, and to secure for the British Empire the honour of this
achievement'. Scott made it to the Pole on 17 January 1912, only to
discover that the Norwegian Roald Amundsen had beaten him by five
weeks. When he died on 29 March trying to make it back to the ship, he
was mourned as a hero and memorialised throughout the empire.[21]

Rival interest in Antarctic territory later prompted the formation of
the British Australian New Zealand Antarctic Research Expedition
(BANZARE), which was instructed to 'plant the British flag whenever
you find it practicable'. True to this imperial mission, the Union Jack was
raised on Proclamation Island when *Discovery* returned to Antarctica with
BANZARE in 1929–31. HMS *Nigeria* was sent to reaffirm Britain's
Antarctic claims in 1948, as public awareness was piqued by the cinematic

release that year of *Scott of the Antarctic*. Today, the country's political and scientific interests continue to be supported in the region by the Royal Navy's Antarctic Patrol Ship, HMS *Protector*.[22]

CHAPTER 3

# Pax Britannica

The defeat of Napoleonic France in 1815 left Britain as the preeminent global power,[iv] ushering in an era until the end of the century which would be known as the *Pax Britannica*, the 'British Peace'. At this time, global trade flourished, slavery and piracy were supressed and liberalism was spread, all due to the supremacy of the Royal Navy which occupied a paramount position across the world's oceans.[v] Yet, despite the nominal tranquillity, it was also a period when the British Empire's borders expanded the most, with the navy often using force to maintain this status quo in Britain's interests. Industrial advantage and technological superiority meant that 'gunboat diplomacy' became the tactic of the age. While this usually brought foreign rulers to the bargaining table to secure favourable terms for Britain, gradually it destabilised indigenous politics and societies to a point where their territorial annexation was considered necessary for safeguarding Britain's commercial and strategic position.

## Suppressing slavery

The abolitionist movement grew from the mid-eighteenth century, as the political influence of plantation interests declined in Britain and the 'sugar islands' came under increased pressure from slave uprisings, shrinking

---

[iv] See also Martin Robson, *A History of the Royal Navy: The Napoleonic Wars*.

[v] See also Andrew Baines, *A History of the Royal Navy: The Victorian Age*.

profits and competition from richer producers in Brazil and Cuba. Free trade sentiments spread amongst British merchants as slave-related traffic decreased proportionally to other imperial commodities, such as palm oil, which met industrialisation's demand for lubricants. A major source of palm oil was in West Africa, and it became increasingly important to Britain's economic interest to cultivate this product over the export of slaves from the region.[1]

A year after the 1807 Slave Trade Act was passed, the Royal Navy established a 'Preventative Squadron' and captured a major slaving port. This became the British crown colony of Freetown and base for the Royal Navy's West Africa Station, and grew as freed slaves settled there to avoid being re-enslaved elsewhere in Africa. Other colonial possessions were important for keeping the squadron at sea, with Cape Town and the mid-Atlantic island of Ascension used as supply depots. While 5.5 million West African slaves were exported between 1701 and 1810, this fell to 1.9 million between 1811 and 1870. In areas such as the Niger Delta and the Western Nigerian coast, where the West Africa Squadron concentrated its efforts, palm oil exports overtook that of slaves, with Liverpool's receipt of 150 tons in 1806 growing to 13,000 tons by 1839. In 1851, the island kingdom of Lagos was occupied and made a British protectorate in order to remove Portuguese slave-traders and 'persuade' Africans to switch to 'legitimate trade' in palm oil production.

By April 1860, the British Consul, George Brand, claimed that Lagos had evolved from 'a haunt of piratical slave dealers' to become 'the seat of a most important and increasing legal trade [ . . . ] the natural entrepôt of an immense country abounding in unlimited resources' and 'the natural [base] of operations for extending the blessings of industry, commerce, and Christian civilization to this portion of [ . . . ] Africa'. He argued that Lagos could 'never fully serve these great purposes under the Native Government', which offered no protection of property and no means of enforcing credit, and therefore 'to do justice to this place', he urged Britain to extend its 'civilized command' over the kingdom. In June 1861, the Foreign Secretary, Lord Russell, issued orders to turn the 'anomalous protectorate into an avowed occupation' that would be 'permanently beneficial to the African race' by supporting legitimate trade and stymieing the ambitions of Ghezo, King of Dahomy, whose 'barbarous

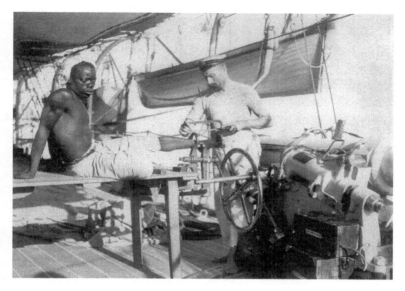

Fig. 3.1. A Royal Navy carpenter saws off a slave's shackles

wars, and encouragement to slave-trading, are the chief cause of disorder in that part of Africa'. Commander Norman B. Bedingfield, Senior Naval Officer of the Bights Division, invited Dosunmu, the *Oba* (King) of Lagos, aboard the sloop HMS *Prometheus* on 30 July 1861 to impress upon him Britain's proposed annexation. Dosunmu promised a reply once he had consulted with his chiefs, but, by 1 August, opposition had grown and he refused to sign the treaty of cession. Bedingfield returned on 5 August, the day before the deadline, backed by 'two brass-gun boats' and 'lots of well-armed marines', and Dosunmu acquiesced with the proviso that he retain the title of 'King' and the power to settle 'native' disputes, subject to British law. On 6 August 1861:

> an immense crowd [ . . . ] collected about the Consulate [ . . . ] and at 1 P.M. the King landed under a salute of seven guns from the 'Prometheus' anchored close by. After signing the Treaty, with four of his principal Chiefs, they were conducted to the flag-staff [ . . . ]; the Proclamation [ceding Lagos to the Queen of Great Britain] was read [ . . . ] the British flag unfurled, and saluted with twenty-guns; the National Anthem sung by a band of children from the Missionary Schools.

The day ended with dinner aboard the *Prometheus* for Dosunmu, his chiefs and nearly all of Lagos' European residents. Though Dosunmu later sent letters of protest to Queen Victoria, claiming he had only complied because Bedingfield threatened to fire upon Lagos, causing its inhabitants to panic and flee, he still received an annual pension of £1,000 from the deal until his death in 1885.[2]

From Lagos, Britain's presence followed an increasing flow of British merchants into the mainland interior, bringing contact with other indigenous states. Local rulers who displayed a willingness to collaborate with the British and imbibe their values of 'Christianity, commerce, and civilization' were allowed to retain their positions, while 'gunboat diplomacy' was used on those who refused to cooperate. After a few demonstrations of the Royal Navy's destructive power, the mere threat of force was often enough to sway dissenters into signing 'unequal treaties', which involved heavy sanctions should they break the agreement.

It also meant that the Royal Navy had to be called in when Britain's collaborative systems were undermined. In December 1857, the Senior Naval Officer in West Africa, Commodore Charles Wise, was ordered to proceed 30 miles north of Sierra Leone up the Great Scarcies river, where he was to punish the Susu, who had attacked Britain's local allies the Timmanees and set fire to a number of British factories. When the Susu refused to leave the Timmanee town of Kambia, the Commodore took his force of eight paddle steamers, including HMS *Vesuvius*, *Pluto*, *Spitfire* and *Ardent*, a rocket cutter, a colonial gunboat carrying the Governor and a party of 250 seamen and marines up the river; within 15 minutes of arriving at Kambia on 1 February, naval rockets had set the town ablaze and 200 Susu were dead from shell bombardment. Wise's squadron only suffered ten wounded men, as they proceeded along the river and destroyed the villages of Robelli, Makanka, Robaiyu and Rokon. The decision not to land an occupation force spurred the Susu on to further aggression, and Wise had to return in March 1859 with 52 boats, a party of marines and the 1st West India Regiment, which inflicted heavy losses on the Susu and succeeded in driving them out of the area before returning it to the Timmanees.

Yet the more Britain had to intervene and assert its authority, the more it weakened indigenous power structures and eroded the authority of local

DURNING OF SLAVE ESTABLISHMENTS, ON SOLYMAN RIVER.

Fig. 3.2. Burning of slave establishments on the Solyman river

rulers in the eyes of their people, increasing the likelihood of further British intervention to protect their economic and strategic interests. Formal colonisation was considered necessary to stem the flow of slaves from civil wars in the Yoruba hinterland of western Nigeria, and to curb French imperial ambitions in the region. A total of 367,928 square miles was added to the empire to meet the growing demands of British enterprises to control the sources and production of raw materials. In this way, the abolition of the slave trade

> cloaked the entire conquest of Africa in a humanitarian guise by presenting European rule and capitalist enterprise, including the employment of freed slaves, as anti-slavery measures. Thus, the ideology of the anti-slavery movement became part and parcel of the European mission to civilize Africa.[3]

Not until the mid-nineteenth century would the Royal Navy's anti-slavery mission extend to East Africa, as popular attention was drawn to the 'Arab slave trade' by the exploits of explorers such as David Livingstone. The deployment of HMS *Lyra* to the region coincided with the appointment of vocal abolitionist, Christopher Rigby, as the British Consul at Zanzibar, and the election of aggressive free trader Lord

Fig. 3.3. Rescued slaves in their new clothes

Palmerston as prime minister. The Royal Navy's initial efforts were
largely ineffectual, as it was used to dealing with large, square-rigged
European vessels. Warships such as the 20-year-old paddle steamer HMS
*Gorgon*, with a speed of nine knots, struggled to keep pace with the
smaller slaving *dhows*, possessing lighter draughts and lateen-rigging, that
could tack against the wind. From 1865, the responsibility fell to the East
Indies Station, which operated less than a dozen vessels spread across the
entire Indian Ocean region and produced only 1,097 slave captures in
1868, less than 5 per cent of East Africa's slave exports.[4]

Special Envoy Sir Bartle Frere and Consul John Kirk, a former
companion of Livingstone, pressured Sultan Barghash into signing a treaty
in 1873 which closed Zanzibar's slave markets, granted the Royal Navy
the right to search Zanzibari vessels, and gave protection to all freed
slaves. To enforce this, Captain George Malcolm suggested stationing the
72-gun HMS *London* at Zanzibar to serve as a depot ship, allowing the
repair and replenishment of several smaller, faster yachts and steam
launches, which emitted less smoke and were more difficult for slavers to
spot from afar.

Fig. 3.4. A captured slaver aboard HMS *Sphinx*

Kirk wanted to increase Britain's influence in East Africa without accruing the additional political and economic costs of direct intervention, using the Sultan as a proxy and strengthening his power so that British subjects could economically develop the region under his patronage. Lieutenant Lloyd Mathews of HMS *London* had seen service during the Second Asante War of 1873–4, and was seconded from August 1877 to establish, train and command the Sultan's army, subsidised by Britain. In December 1881, Mathews led this force in capturing Hindi bin Hattam on Pemba Island after the slaver had killed *London*'s Captain Charles Brownrigg during a boarding action to free 100 slaves. Mathews' army tightened control over the Sultan's mainland subjects beyond the slave trade, providing a buffer against Britain's European rivals looking to annex the Sultan's territory. The 1870s and 1880s was a period of 'new imperialism' and growing European interest in Africa. In November 1884, a conference was held in Berlin, where the continent was carved up between the major powers. As in West Africa, imperial ambitions were masked beneath humanitarian guises, with a commitment to ending the Arab slave trade included to attract popular support for the agreement. The Royal Navy's efforts had achieved some success, just a month before the conference, HMS *Philomel* captured two *dhows* containing over 200 slaves off the coast of Oman. When Ali bin Said became Sultan in February 1890, around 150 Germans represented the largest European community in Zanzibar. Britain used Sultan Ali's fear of German encroachment to convince him to formally seek its protection. A treaty in July that year established Zanzibar as a British protectorate in exchange for Germany receiving the North Sea island of Heligoland. Mathews became First Minister, reporting only to the Sultan and the British Consul, while British officials headed other government departments.[5]

From this position, Mathews was able to act in Britain's regional interests. He used a force of 300 bluejackets and 200 Zanzibari troops in June 1895 to quell a rebellion in the town of Takaungu – between Mombasa and Malindi – which had risen up against Britain's nominated local ruler, Sheik Rashid bin Salim. Upon Sultan Hamid bin Thuwaini's death in 1896, his cousin, Khalid bin Barghash, staged a *coup d'état* against the pro-British Hamoud bin Mohammed. At 09:02, on 27 August, Mathews ordered Royal Navy warships stationed near Zanzibar to bombard the palace. By 09:40, the palace had been destroyed, killing 500 of Khalid's followers and forcing his surrender. The 'war' lasted just 38 minutes – the shortest in history. It was a classic case of gunboat diplomacy, or, in modern parlance, 'shock and awe', with the Royal Navy's technological superiority upholding Britain's interests in a demonstration of imperial power that secured local allegiance and dissuaded opposition. Upon his death on 14 October 1901, Mathews was eulogised as a

> soldier and sailor and vizier, he was of the fibre of those simple, God-fearing mariners of the great days who laid the foundations of empire [ . . . ] no one has been ever more sincerely mourned by those of an alien race and dusky skin.

While Mathews had succeeded in establishing British imperial influence in Zanzibar, thousands of slaves were still being imported through the Omani port of Sur. Though the African slave trade decreased, as late as 1922, HMS *Espiegle* shelled the town of Khadhrah, near Suwaiq, to dissuade local residents from harbouring enslaved Baluchis.[6]

Fig. 3.5 View of Bermuda with floating dock

The Royal Navy also materially benefitted from its anti-slavery mission. When the American Revolution broke out in 1775,[vi] the British sought ways of increasing their manpower on the continent through alliances with the Native American and African-American populations, while encouraging slaves to rebel against their masters in exchange for freedom. Britain's defeat and the loss of its American colonies meant that thousands of African-American soldiers had to be transported to Britain, finding employment in the Royal Navy, Merchant Navy and British Army. The loss of harbours to the United States also meant that ex-slave labour was employed in developing the naval yards at Halifax, Nova Scotia (first opened in 1759), Kingston, Ontario (from 1788) and Bermuda (from 1795).

African-Americans recruited into the navy would serve as far afield as India. Such men included William Hall, who was born in Nova Scotia in 1827 as the son of a freed slave. Hall served with the Naval Brigade from HMS *Shannon* during the Indian Rebellion, when he became the first black man to receive the Victoria Cross for his part in relieving the British Residency at Lucknow in November 1857. He would go on to become a

---

[vi] See also Martin Robson, *A History of the Royal Navy: The American Revolutionary War.*

petty officer and quartermaster of HMS *Peterel*. While Black and Asian seamen represented just 5 per cent of the Royal Navy's Indian Ocean personnel in 1861, within 20 years this figure would more than double to 11.55 per cent, as its anti-slavery role expanded in the region.[7]

Since the eighteenth century, the 'Kru tribe' of Liberia had been recruited as sailors, cooks and interpreters aboard European vessels operating along the West African coast, and they developed a reputation as especially skilled seafarers. They were physically marked out from other Africans by a vertical line tattooed down the centre of the forehead; this was believed to represent a ship's mast and it ensured that they were not enslaved. Originally employed aboard slaving vessels, they would later serve with the Royal Navy's Preventative Squadron. In the 1861 census, 'Kroomen' appeared aboard HMS *Espoir* in West Africa, HMS *Sharpshooter* off the Congo and HMS *Persian* in South Africa, while 21 Kru travelled as far as Hong Kong with HMS *Algerine*. One hundred and eighteen Kru were also employed as naval labourers on Ascension Island. Favourable reports from Rear Admiral Walker of the Cape Station meant that Royal Navy ships en route to South Africa would stop at Sierra Leone to contract Kroomen, and, by 1863, almost 100 Kru were deployed aboard its vessels in East Africa.[8]

When the Royal Navy captured a slaving ship, its cargo of 'Prize Negroes' would be 'liberated' at Cape Town, where they would be 'apprenticed' into a system of indentured servitude for 14 years. Some were granted apprenticeships with the Royal Navy, while those hired elsewhere provided an important revenue stream for the service. As women were also liberated, the Royal Navy contributed to the formation of new communities of ex-slaves around ports adjoining the Atlantic and Indian Oceans: from Ascension Island, Cape Town, and Durban, to Aden, Karachi and Bombay. Freedmen liberated and employed by the Royal Navy in East Africa, the Seychelles and Bombay became known as 'Seedies'. This name originated from the nineteenth-century Anglo-Indian English term 'Sidis', representing Muslim mariners from the Swahili coast who were subjects of Zanzibar's Sultan or *Seyyid*. Many Seedies earned a reputation from European explorers as local guides and interpreters on African expeditions. Royal Navy crews would give names to Freedmen they liberated; these names were often connected with the duty they performed, the British royal family or the name of the ship. One example was the Zanzibari, Tom Nimble, who was freed by HMS *Nimble*. From 7 April 1870, the expense of ferrying Kroomen to East Africa and back led the Admiralty to replace them with Seedies on that coast. Opposition from Cape Station officers, who believed that it took 12 Seedies to do the job of eight Kroomen, meant that Kru were still serving in the region during the First Zulu War of 1879 and the First Boer War of 1880–1, with 14 individuals listed on a plaque laid at Simon's Town: 'In memory of the officers and men of the Naval Brigade from H.M.Ships "Boadicea" and "Dido" who fell at Laing's Nek and Majuba Hill on the 28 of January and 27 February 1881'.[9]

Following HMS *London*'s deployment to Zanzibar, the Seedie allowance was increased from 18 to 33. Its commanding officer, George Sullivan, had valued Zanzibari interpreters since serving in the Indian Ocean during the 1840s, and they played an essential role in the anti-slavery campaign since few Royal Navy officers could speak Kiswahili. By 1881, ten Royal Navy ships and 1,245 personnel were serving in East Africa; they included 56 Asian Lascars drawn primarily from Goa, and 153 Seedies from Africa and Arabian ports including Aden, Muscat and Jeddah. African stokers also emerged, such as HMS *Euryalus*' Jack Daphne, who was likely freed by HMS *Daphne* when it captured a *dhow* containing 225 slaves on 28 February 1874.[10]

The Royal Navy's remit stretched far from the sea to the great lakes of Africa's interior. In 1895, the British Commissioner of the Central African Protectorate (present-day Malawi) and Lieutenant Percy Cullen, Royal Naval Reserve, led a force of 300 Africans, 100 Sikhs, a Naval Brigade and the gunboats *Adventure, Pioneer* and *Dove* in operations against the Arab slave-traders Kawinga, Matapwiri, Zarafi, Mponda, Makanjira and the self-styled 'Sultan' Mlozi, who had established a fort at Karonga. To make an example of him, Mlozi was captured and hanged for his 'crimes and atrocities' in front of the Wankonde chiefs. Furthermore, an administration station and fort were erected near the British South Africa Company's boundary to protect against Awemba raids.[11]

The Royal Navy's recruitment of non-European recruits stalled following the 1901 War Office declaration that the 'cardinal principle of British policy' was that 'African armies should not be used in "white men's

Fig. 3.6. Crew members of HMS *Rattler, c.*1901–3

wars"' and that the main burden of imperial defence 'must be borne by
the white subjects of the King'. In the midst of the Boer War, the use of
black Africans by both sides had aroused a fear of empowered bodies of
'armed and disciplined blacks disabused of the sanctity and solidarity of
the white man by their wartime experiences' rising up to challenge
European dominance. Once slavery was abolished within the Zanzibar
Sultanate in 1907, and with regulations introduced in 1906 stating that
naval 'candidates must be of pure European descent', East African
recruitment ceased until the manpower shortages of World War I. Lascars
were present during the battle of Zanzibar, when the German cruiser SMS
*Königsberg* sunk HMS *Pegasus* on 20 September 1914, while Somali
crewmen served with HMS *Venus* at Singapore in 1916. As the number of
Swahili recruits declined and were replaced by men from British
Somaliland and Aden, the term 'Seedie' fell out of Royal Navy usage and
became superseded by a more general definition of 'Somali' on 14 May
1934.[12]

Slavery was also prevalent amongst the First Nations of north-west
America until the 1860s, with one in 17 Aboriginal people west of the
Rocky Mountains enslaved in 1845. The 'Great Game' with Russia and
the 'manifest destiny' of the United States' western expansion influenced

Britain's colonisation of the region. James Douglas, the mixed-race Governor of Vancouver Island from 1851 and British Columbia from 1858, believed that in order to establish favourable commercial and political conditions for British settlement and the Hudson's Bay Company, who he had previously worked for, slavery needed to be suppressed by extending Britain's 'moral influence' with Royal Navy assistance. In July 1860, Captain Richards, commanding HMS *Plumper*, *Termagant* and *Alert*, was despatched by Douglas to extract promises from the neighbouring tribes that they would stop warring with one another and live peacefully according to British laws. Here, gunboat diplomacy morphed into 'forest diplomacy', with naval captains conducting conspicuous live fire exercises while visiting indigenous villages. At Laskeek, in the Queen Charlotte Islands, Chief Konyil of the Haida was 'made visibly nervous' by the presence of HMS *Alert*, and agreed to end his people's slaving raids. Similar assurances were gained from the Tsimshian and Kwakiutl, and the decline in inter-tribal violence limited opportunities for taking and trading slaves.

Naval officers also targeted 'unscrupulous traffickers' of alcohol, who contributed to the problem by inciting aggression in normally 'quiet and inoffensive' people. While HMS *Reindeer*'s Captain W.R. Kennedy believed that the indigenous population were 'a very degraded race', he conceded that they were corrupted by 'the towns, where civilization in the form of whisky and disease have been brought to bear on them', though some brewed their own fermented drink 'hoochinoo'. Governor Douglas directed Commander John W. Pike and HMS *Devastation* to check the illegal traffic in alcohol, and in September 1862 they destroyed 300 gallons of spirits from the schooner *Hamley*, discovered near Dundas Island. Further seizures were made the following year at Hornby Island (300–400 gallons of spirits), Fort Rupert (70 gallons), Bishops Cove (300 gallons) and Nass river, while the floating distillery *Langley* and vessel *Petrel* were impounded and sold along with their cargo, raising £201 9s. 6d in prize money for *Devastation*'s crew. Royal Navy patrols against the liquor trade continued well into the 1870s.[13]

Britain's 'pacification' of North America's First Nations, and indigenous groups elsewhere, inevitably relied upon it possessing a monopoly of violence. While military clashes between Aboriginal people and Europeans further into the interior of the country were much more evenly balanced, on the coast or along rivers the fire support provided by mobile naval guns and professional marines proved decisive. Consequently, the royal proclamation of 1763 and land cession treaties signed with the interior tribes never extended to the coast where the British occupied a stronger position. They still faced local resistance, particularly following the influx of British fishermen into traditional fishing grounds in the east; in 1715, the Mi'kmaqs held English fishing boats to ransom in Cape Sable, destroyed the Canso fishing station in 1720 and captured 18 vessels in July 1722, the latter requiring the intervention of two Royal Navy sloops. In the Miramichi river, HMS *Viper* apprehended 16 Mi'kmaqs after they attacked a trading post, while Mi'kmaq and Maliseet attempts to unite against the British were dissuaded by HMS *Vulture*'s presence in the St John river. When the Newitty murdered three British deserters near Fort Rupert in 1850, a corvette was despatched to arrest the culprits, but the landing party's response – setting fire to the village and dispossessing its inhabitants – was vengeful. Similarly, when the British merchant ship *Kingfisher* was captured and her crew were killed at Clayoquot Sound in 1864, the *Devastation* and HMS *Sutlej* destroyed nine villages, 64 canoes, and killed 13 Aboriginal people in return. Such punitive expeditions continued in Canada until the 1880s.[14]

## Gunboat diplomacy: The Opium Wars

Lord Palmerston's foreign policy opened up new regions of the world to British commerce and investment by applying naval and diplomatic pressure to secure free trade agreements. Warships were more than merely a means of projecting physical military power; they also had a psychological impact, being the manifestation of Britain's industrial, technological and economic supremacy. While English sailing vessels had opened fire on Chinese forts guarding the Pearl river in 1637, their lack of manoeuvrability meant that they struggled to traverse the river and were vulnerable to groundings and close cannon fire from shore. For China, the British remained 'barbarians from the sea', an annoyance along the coast but one which posed little threat to its major cities inland. The development of the steam 'gunboat' changed that, and gave Britain the ability to project its influence up-river into the Chinese interior, bringing valuable centres such as Canton into its reach. 'Gunboat diplomacy' was built upon a principle of 'carrot and stick'. For the China Station's Captain, Henry Keppel, it was important to impress the locals with demonstrations of the Royal Navy's superior training and discipline, thus upon approaching port his 'ships broke their masthead flags, manned yards, and fired royal salutes; looking as no other nation's ships can look – ropes taut and yards square'. When a country was asked to sign a treaty with Britain in the presence of a Royal Navy warship, the message it conveyed was both seductive and threatening; while it offered Britain's friendship and the Royal Navy's protection from one's enemies it also showed that, should a ruler refuse to do business, the Royal Navy gave Britain the power to take what it wanted by force. More often than not a country would acquiesce, though China famously resisted and suffered dire consequences in what would be known as the 'Opium Wars'.[15]

No event better demonstrates both the interconnectedness of British economic and maritime strategy at this time and the naval dominance that allowed it to achieve those goals, than Britain's two wars with China, from 1839 to 1842 and 1856 to 1860. For centuries, Europe had been trying to open up China's potentially huge market to trade. While, by the eighteenth century, Britain was importing large amounts of tea, silk and porcelain from the Celestial Empire, its exports in return were virtually non-existent. China believed it was the centre of the world and culturally superior to the 'western Barbarians', who it saw as offering nothing of value that the Chinese could not better produce themselves. Consequently, Britain suffered a large trade deficit, with money flooding out of the country to pay for Chinese commodities and little flowing back in the opposite direction.

In order to redress this imbalance, under the auspices of the British East India Company, Britain started exporting opium from Bengal to China in 1773, with an estimated 1,000 chests arriving in Canton each year. By the 1820s, opium had provided Britain with a trade surplus, with shipments more than doubling by 1840:

The Daoguang Emperor, concerned about the drain of wealth from his country but also the social and moral ills the addiction had wreaked among the Chinese people, charged Lin Zexu with eliminating the opium

**Table 3.1. Opium imports to China from India**

(1 chest = £140 approx.)

| Year | Imports |
|------|---------|
| 1773 | 1,000 chests |
| 1790 | 4,000 chests |
| Early 1820s | 10,000 chests |
| 1828 | 18,000 chests |
| 1839 | 40,000 chests |
| 1865 | 76,000 chests |
| 1884 | 81,000 chests (peak) |

trade. In June 1839, he confiscated and destroyed over 20,000 chests of opium worth approximately £2.5 million. In response, a British expeditionary force was dispatched to blockade the Pearl river. When, on 3 November, 29 Chinese war junks were sent out to engage the British force the 22-gun, sixth rate HMS *Volage* and the 18-gun sloop HMS *Hyacinth* comprehensively routed them:

> One of the war-junks was now on the beam of the Volage, and fired a couple of guns at her, which passed over. These were immediately returned, several of the shot telling on the junk; and almost instantly we heard an explosion, and on looking round saw through the envelope of the smoke the fragments of the unfortunate junk floating as it were in the air. She had blown up. When the smoke cleared somewhat off, out of whatever number she might have had on board, we could see but three about the wreck [ . . . ] not being fitted for elevation or depression, all their shot were too high to have any effect, except on the spars and rigging.[16]

This signalled the start of hostilities and set the tone for what would follow. A most striking example of the 'Great Divergence' between China and the West was the industrial–technological disparity between the fast steam-driven, iron-clad warships of Britain's naval forces, and the wooden sailing junks of the Chinese Navy:

> In some instances – as in the attack on the Bogue forts or on the city of Tinghai – steamers were used as tugboats to pull the big ships of the line into position to fire their broadsides at the enemy. At other times they pulled boats full of

marines to the site of amphibious attacks; shallow-draft steamers [ . . . ] were especially suited to such operations. They were also very effective against another favourite Chinese river tactic: fireboats filled with oily cotton set ablaze and cast adrift to smash against the British men-of-war. The steamers simply grappled them with hooks and pulled them out of reach of the warships. With their quick maneuvers and Congreve rockets, river steamers could sink the best Chinese war junks with no trouble.[17]

The military and psychological devastation wrought by Britain's naval advantage was epitomised by the 184-foot, 660-ton iron steamer *Nemesis* during the second battle of Chuenpee. Originally built for the British East India Company, but commanded by Commander William Hutcheon Hall of the Royal Navy and operated under an Admiralty letter of marque, the *Nemesis* combined a number of technological innovations that made it a fearsome weapon during the First Opium War. Made almost entirely out of iron, it was driven by two paddle wheels, while its flat-bottomed hull gave it a draught of just 6 feet when fully laden, allowing it to navigate China's shallow waters with ease. On 7 January 1841, the *Nemesis* bombarded the fortifications at Chuenpee, landed a force of 600 Indian sepoys, and destroyed the Chinese fleet of war junks in a morale-sapping demonstration of 'shock and awe' that earned it the name 'devil ship' among the Chinese. As Commander Hall recalled:

the Nemesis, having the great advantage of drawing less than six feet water, was able to approach near enough to bring her two 32-pounder pivot-guns to bear within good range [ . . . ] The very first rocket fired from the Nemesis was seen to enter the large junk against which it was directed, near that of the admiral, and almost the instant afterwards blew up with a terrific explosion, launching into eternity every soul on board, and pouring forth its blaze like the mighty rush of fire from a volcano. The instantaneous destruction of the huge body seemed appalling to both sides engaged. The smoke, and flame, and thunder of the explosion, with the broken fragments falling round, and even portions of disseevered bodies scattering as they fell, were enough to strike with awe, if not with fear, the stoutest heart that looked upon it.[18]

By 29 August 1842, Britain had forced China's capitulation, and the first of the so-called 'unequal treaties' was signed aboard HMS *Cornwallis*. In addition to paying war reparations and compensation for the destroyed

Fig. 3.7. Signing and sealing of the Treaty of Nanking aboard HMS *Cornwallis*

opium, the Treaty of Nanking forced China into opening up five treaty ports to foreign trade, limiting import duties to 5 per cent and ceding Hong Kong to Britain. By 1850, the duties raised from Chinese tea alone more than covered the expense of maintaining the Royal Navy.

The opium trade was legalised by the Treaty of Tientsin, which followed the Second Opium War of 1856–60. In that conflict, the Royal Navy had used over 25 gunboats to attack Canton and the Taku forts near Peking, securing for Britain additional treaty ports, the cession of the Kowloon peninsula and freedom of movement for Christian missionaries in China.[19]

Though the gunboat no doubt proved its military effectiveness during the Opium Wars, its so-called 'diplomacy' as a deterrent was much less successful in this case. 'Gunboat diplomacy' is to a large extent a psychological strategy, combining naval prestige with economic or political incentives, and a threat strong enough to dissuade resistance without actually resorting to military action. The overwhelming technological superiority of the Royal Navy could not overcome Sino-centric attitudes regarding the West's cultural inferiority to stop the Opium Wars from breaking out. But to prevent a loss of prestige elsewhere, which might create a domino effect of opposition, Britain had to live up to its latent threat on occasion. Despite Chinese defiance, Britain's reputation was enhanced by demonstrating the power that the Royal Navy could bring to bear anywhere in the world, to friends and enemies alike. The 'carrot and stick' of the service was used successfully elsewhere, such as Morocco in 1858, after the country reneged on its free trade treaty when Britain's attention was diverted by the Crimean War:

Fig. 3.8. Bluejackets in the Battle of Canton, December 1857

It was pointed out [to the Sultan] that Turkey, another Mohammedan state, had concluded, at the desire of the British government, the treaty of 1838, which had caused a great expansion of trade between the two countries: so at that moment Great Britain was sending 'large fleets to support the Turks, who had had unjust demands made upon them by the Russian government'. Such, it was intimated, might be the good fortune of Morocco in similar moments of difficulty if she followed Turkey's example in fostering commercial relations with Great Britain. On the other hand, 'if the Sultan rejected British counsels regarding commercial matters he could not expect British support in moments of difficulty as previously' [ . . . ] when the drafts were referred to the Sultan's ministers, they were so emasculated as to be virtually useless. Hay met this last obstruction of the monopolists by [ . . . ] a declaration of his intention to go to the various ports on a British warship and exact from the local authorities on behalf of the British traders the benefits conferred by the treaties. This threat, combined with reminders of the magnificence of the presents to be made on the completion of negotiations, served to bring about the conclusion of the new agreements.[20]

## The imperialism of free trade: South America

The disintegration of old imperial rivals gave Britain an opportunity to fill the power vacuum left behind. Whereas Spain's empire was essentially a parasitic one, draining its colonial possessions of their natural resources to enrich the Iberian Peninsula, this imperial model was inherently unsustainable once the value of gold began to decline. Though the early English Empire emerged from similar origins, represented by privateers such as Drake, by the nineteenth century Britain pursued an 'imperialism of free trade'. This was based on a notion of acquiring the commercial benefits of empire without accruing the administrative costs of territorial annexation. With the support of gunboat diplomacy, this could be achieved by establishing free trade agreements with otherwise sovereign states, bringing them into Britain's economic and political sphere of influence, or its 'informal empire'. Foreign Secretary George Canning envisaged this when he declared in 1824 that: 'Spanish America is free and if we do not mismanage our affairs sadly, she is English'.[21]

Britain had long sought to profit from the self-determination of Spain's South American colonies and supported their struggle towards it. Minister for War, Lord Castlereagh advocated in 1807 that 'the liberation of South America, must be accomplished through the wishes and exertions of the inhabitants; but the change can only be operated [ . . . ] under the protection and with the support of an auxiliary British force'. After being proclaimed El Libertador ('the liberator') of Venezuela in 1813, Simon Bolivar conceded that 'only England, mistress of the seas, can protect us against the unified force of European reaction'. With Britain and Spain at war, the Royal Navy blockaded Cadiz, which allowed British merchants to make up the shortfall in Spanish colonial trade and meant that the value of Jamaica's exports rose to over £1 million in 1808. Two years later, the Royal Navy helped evacuate the Portuguese royal family to Brazil. This gave Britain leverage to extract a trade treaty granting it a lower import tariff than Portugal, allowing its manufactured goods to flood Brazil's market, while the export of Brazilian sugar and tobacco remained prohibited to protect British West Indian producers. While temporarily discharged from the Royal Navy, Lord Thomas Cochrane played an important role in South America's liberation, taking command of Chile's naval force in 1817 during its revolution against Spain. In 1820 he led the Freedom Expedition of Peru, contributing to that country's independence, before aiding Dom Pedro in Brazil's secession from Portugal. During the summer of 1835, Commander Sir James Everard

Home and HMS *Racehorse* helped solidify Brazil's post-colonial regime by laying siege to the rebel town of Para, while Mates Baldwin, Wake and Drury landed ashore and extracted 220 fugitives from among the insurgents.[22]

In Argentina, the expansion of British trade saw the coastal regions flourish at the expense of the hinterlands, contributing to the collapse of that country's constitution and the eruption of civil war. General Rosas seized power and was less cooperative, having designs on Montevideo which closed the Rio de la Plata off to trade. Britain had supported Uruguay as a buffer state since the 1820s, and, with the French, it sent a naval force under the command of Captain Charles Hotham to reopen the River Plate. On 20 November 1845, the Argentine fortifications at Vuelta del Obligado in the Parana were destroyed. The naval presence along the river restored confidence in Montevideo's economy and allowed trade to resume, whilst it isolated Rosas' army in Uruguay and stirred up local opposition against Buenos Aires. Displays of 'naval theatre' won local 'hearts and minds' over to the British; on 25 February 1846, the paddle sloop HMS *Alecto* 'showed the flag' at Corrientes, drawing out the whole community who cheered the rare appearance of a modern steamer:

> Crowds lined the banks and in the town every rooftop and window was crammed with spectators. 'As soon as the heat of the day had passed, the people began to swarm on board and speedily the whole vessel was thronged' [ . . . ] the gun room was opened to visitors and cherry brandy was served to all. Ashore in the town again, [sailors were] 'frequently invited into houses and regaled with *mate* and cigars'.

By 4 June 1846, when the warships left to engage the new Argentine batteries at San Lorenzo, 110 merchantmen were once again operating along the River Plate. Captain Hotham justified the naval intervention by arguing that 'a trade which gave full employment to 140 or 150 merchant vessels during the few months of the English and French occupation is not to be despised', and in 1852 the major tributaries of the River Plate were recognised as international waterways by Argentina.[23]

South America's importance to Britain grew with the development of its antipodean colonies. Australian wood and oil, New Zealand flax and timber, and exotic goods from Polynesia, were transported to Britain around Cape Horn. From 1822, Britain also imported copper ore from Peru and Chile, whale products from the North Pacific, guano from the Chinca Islands off Peru and silver from Mexico, while British exports to Central and South America grew by 93 per cent between 1816 and 1842. As Britain's economic interests in the region increased, so did the strategic value of a Royal Navy station at the Falkland Islands,

> which from its locality becomes daily more requisite, in consequence of the increasing extension of our Australian and South American commerce, and the absence of any other naval position belonging to Great Britain, nearer to the southern hemisphere than the Cape of Good Hope [ . . . ] Placed intermediately between both seas, with a harbour of refuge at hand to refit in, and a good lock-out, it would be difficult, if not impracticable, for any ship to navigate those seas without the permission of Great Britain.

The first documented landing on the Falkland Islands was by the English Captain John Strong in 1690, and a British settlement existed there until 1774. From 1783, Britain identified the Falklands as a potential base for its whalers and sealers, and the Admiralty pushed for the islands' re-occupation. Supporting this, the Permanent Secretary of the Admiralty, Sir John Barrow, advised the Colonial Office in 1829 that the 'law of nations' dictated that 'priority of discovery must give way to priority of occupancy'. That year, Argentina named Luis Vernet as the military and civil commander of the islands, but his attempts to end foreign sealing drew the ire of the USS *Lexington*, which destroyed the nascent settlement in December 1831. To thwart further American and Argentine designs, Palmerston instructed the Admiralty to reassert Britain's sovereignty in August 1832, and Commander John James Onslow was dispatched from the South American Station with HMS *Clio*. Though an Argentine garrison and schooner, *Sarandi*, had already returned to East Falkland, the *Clio*'s arrival in December prompted their mutiny and allowed Britain to re-establish its presence on the islands.[24]

### 'Pacific' missions

In many parts of the world, Royal Navy captains were Britain's most senior representative, serving as both the voice and arm of its authority and entrusted with consular powers allowing them to intervene in matters concerning British subjects. This was particularly true of the vast Pacific Ocean, patrolled by no more than 20 older, smaller and slower vessels, based out of Sydney and Valparaiso in Chile, with over 7,000 miles of ocean between them. In 1829, Commander Laws and HMS *Satellite* visited the Society and Tongan Islands, and recommended that an annual cruise be established there 'to countenance and encourage the British intercourse and Trade'. While Royal Navy captains were urged to foster 'that amicable intercourse with the Native Chiefs and peoples which their growing commercial importance requires', they held concerns about the 'great numbers of worthless characters calling themselves Englishmen, from New South Wales and elsewhere, who keep the natives in constant dread by their depredations'. It was feared that the behaviour of these white men might undermine British authority, and five Englishmen were arrested by HMS *Seringapatam* during Captain Waldegrave's cruise of 1830 because their poor example inspired a 'native bias against England'.

Christian missionaries were described by one Royal Navy Captain as 'avant couriers of commerce', who required 'protection as well as watching'. Imperial rivalry could incite religious sectarianism among indigenous converts, and, in September 1844, Captain Home of HMS *North Star* had to quell Wallis Islanders after the Protestant church was destroyed by local Catholics, who were convinced by the bishop that French warships would punish them if they did not obey his command.

Upon arriving at Apia in February 1847, Captain Blake of the Pacific Station was asked by Samoa's Consul Pritchard to use his firepower to settle a local dispute. He refused, however, not wishing to compromise what he believed was the Royal Navy's higher paternalistic duty:

> As the natives, either from fear or terror, entertain such exalted notions of a man of war, I thought nothing could have a more pernicious effect than to make threats which I was not fully prepared to carry out [ . . . ] there was something due from us to a half savage uninstructed people, such as Justice, Charity and moderation, especially on the part of ourselves as a people claiming a superiority not merely physical, but derived from those higher attributes and valuable distinctions and advantages of civilization and refinement, and boasting, perhaps justly, of our pride in belonging to a country whose fame and

whose power is spread throughout the world and whose justice, if I am conscious of it, in the station I hold, I will not be a willing instrument to sully.

The Admiralty discouraged its captains from resorting to gunfire when resolving 'native' disputes, because it would 'defeat the maintenance of those amicable terms of commercial intercourse, which the presence of a Naval Force is intended to cherish and protect'. Blake recognised this, but other captains were unsure of the correct protocol in areas beyond Britain's jurisdiction. Pritchard argued that inaction would undermine the Royal Navy's prestige and Samoan respect for the service, making life difficult for Apia's European residents and Britain's ability to exert influence in the future. While this failed to sway Blake, Captain Worth of HMS *Calypso*, upon receiving Pritchard's petition, destroyed the local village and killed ten of its inhabitants after negotiations failed.

A general absence of British civil authority meant that, in 1854, it was still stressed to Royal Navy captains in the region that:

> it will be your object to give to the natives an impression of the power and of the friendly disposition of the British nation and whilst giving due weight to the representatives of the British consuls and missionaries and to strengthen their hands for good, you will repress any tendency to undue interference or encroachments on the rights of the chiefs and natives.[25]

Usually geopolitical considerations outweighed paternalistic sentiments. A Samoan request for British protection was rejected in 1859 by the Colonial Office because 'the meaning of protection would be, I suppose, that we should be ready to quarrel with any of the first rate maritime states for the sake of the wild inhabitants of these out of the way islands'. If Britain's commercial and strategic position was threatened, however, then the Royal Navy's involvement was more warranted. Naval officers intervened on three occasions to prevent Fiji from descending into civil war. Missionaries praised the firm tactic adopted in 1865 by Sir William Saltonstall Wiseman, commander of the frigate HMS *Curacao*, after one of their own had been expelled from Tana. A veteran of New Zealand's Taranaki Wars, Wiseman demanded £1,000 in compensation from the villagers, and when they refused, fired on the village and destroyed the plantations. Wiseman collected souvenirs from his 'visits', and he exhibited over 1,200 of these 'curiosities' at the Diocesan Book Repository in Sydney during November 1865, including 87 photographs of islanders, missionaries and buildings. Such objectification helped justify the violent action taken against indigenous people, by racially and culturally lowering them in the eyes of Europeans.

The enforcement of fines and punishments increasingly became a feature of the Royal Navy's island cruises, with the Foreign Office considering it to be 'the only mode [ . . . ] of safeguarding the persons and property of British traders'. By 1873, the Fijian government was openly requesting that Britain take formal control of the islands, and the Australia Station's Commodore Goodenough was despatched in November to investigate the prospect. He reported that Fiji was already a virtual protectorate, as for several months tax collection had only been maintained with the assistance of a British warship. As the islands also provided an ideal coaling station between the Australian colonies and Canada, Fiji formally became a Crown colony on 10 October 1874.[26]

## New Zealand: The Maori Wars

Where formal colonisation took place, the experience of naval officers in dealing with indigenous cultures meant that many were appointed to serve as governors. New Zealand's first Governor was Captain William Hobson, who first visited the islands in HMS *Rattlesnake* during the Northern War of 1837, before returning in January 1840 with HMS *Herald* to colonise them for Britain. He negotiated the Treaty of Waitangi with the Maori, granting them exclusive possession of their lands so long as they accepted Queen Victoria's hegemony and the Crown's right to any tract that came on the market. Hobson also set up a naval depot in Devonport, Auckland, which would grow to become New Zealand's main naval base. When he died on 10 September 1842, Hobson was succeeded by Captain Robert FitzRoy, who had earlier commanded HMS *Beagle* during Darwin's voyage.

As some Maori sold their land in exchange for items such as guns and blankets, they felt resentment at being cheated by European traders. This encroachment was symbolised by the flying of the Union Jack on Flagstaff Hill at Kororareka, where the colours of the United Tribes of New Zealand once flew. The British flag was subsequently cut down by the Maori, who sought to regain by force what they had lost. To quell the uprising, in December 1845, Captain Charles Graham led a Naval Brigade of 340 men from HMS *Castor*, *Calliope*, *North Star*, *Racehorse* and the East Indiaman *Elphinstone*, that helped capture Ruapekapeka on 11 January 1846. A party of *Calliope*'s seamen and marines followed up on this by capturing one of the main warring chiefs in South Island, Te Rauperaha, on 23 July 1846.

While the Flagstaff War was brought to an end in 1846 with the use of no more than 1,300 troops, the Taranaki Wars of the 1860s would draw in 4,500 men and 13 warships. Land continued to be a source of antagonism, with the Crown's monopoly forcing the Maori to sell below market value. When rumours of Maori disquiet at a land deal in Taranaki province grew, the Governor requested a warship be despatched to act as a deterrent. Captain Cracroft of HMS *Niger* was less convinced, however, of gunboat diplomacy's effectiveness against a people who were willing to defend their tribal inheritance to the last man. *Niger*'s First Lieutenant Hans Blake took 50 men and a 12pdr howitzer to reinforce the garrison of New Plymouth, but by the time Cracroft returned on 25 March 1860, tensions had boiled over, resulting in the death of a marine and a chest wound for Blake. Cracroft led a successful assault on the Maori *pā* (fortification) on 28 March, in which only four of his men suffered injuries, but celebrations were short-lived as the Maori response required additional Naval Brigades to be landed. A defeat at Puketakauere on 27 June 1860 left HMS *Pelorus*' Captain Beauchamp Seymour wounded, and New Plymouth was reinforced by 150 men and two 8-inch guns from HMS *Iris*. Though peace was agreed in April 1861, Cracroft warned against the Maoris keeping their arms, on the grounds that it would allow them to build another *pā* and resume fighting at a time of their choosing.

Flying the flag of Commodore William Farquharson Burnett CB, Commander in Chief of the Australia Station, the modern screw-driven, steam corvette HMS *Orpheus* was sent to New Zealand in January 1863 to impress the Maori and relieve the sloops HMS *Miranda* and *Harrier*. She would gain infamy in 'New Zealand's worst sea disaster' when, on 7 February, she became wrecked on the sand bars at the mouth of Manukau Harbour, causing 189 of her 259 men to drown, including Commodore Burnett. Most of the survivors had to be rescued by Maori oarsmen from the nearby steamer *Wonga Wonga*. While Burnett had been navigating with an outdated chart, the Admiralty refused to find him guilty in their inquiries that followed and preferred to lay the blame on the colonial authorities and local signalman Edward Wing, though Wing had reported a need for better signalling facilities.

War erupted again just a few months later, when British forces invaded Waikato on 17 July 1863, after the Kingite Maori rejected an ultimatum to swear an oath of allegiance to Queen Victoria or be expelled south of the Waikato river. The success of Britain's new mobile strategy relied upon logistical support from a flotilla of flat-bottomed barges operated by the men of HMS *Harrier*, which allowed operations to be conducted further up the Waikato river. They were accompanied by half a dozen gunboats, whose armour allowed them to conduct close reconnaissance, including the *Pioneer*, which was built in Britain's New South Wales colony and towed across the Tasman Straits by HMS *Eclipse*. Command of the river allowed the British to capture the key Maori *pā* at Meremere on 1 November, contributing to their victory at the battle of Rangiriri three weeks later. The Waikato Naval Brigade was redeployed to Tauranga Harbour on the North Island's east coast, where over half of its number – including Captain Hamilton of HMS *Esk* –were lost in the assault of Gate *pā* on 29 April 1864. *Harrier*'s Commander, Edward Hay, was also mortally wounded, though his Coxswain, Samuel Mitchell, disregarded his personal safety to carry Hay to shelter. For his heroism, Mitchell received the first Victoria Cross to be awarded in Australia, presented by New South Wales' Governor in front of a crowd of 9,000 people at Sydney's Hyde Park. Gate *pā* would be the navy's last engagement in New Zealand, and by 4 May the brigades had re-joined their ships, though cases of fighting continued between Maori and the settlers for another eight years. Upon their return, the surviving officers and men of HMS *Curacoa*, *Miranda*, *Harrier*, *Esk* and *Eclipse* erected an obelisk in honour of their fallen comrades, which still stands today at the former Royal Naval College in Greenwich.[27]

## The Indian Ocean

With the ending of the East India Company's trading monopoly in India (1814) and China (1833), the Royal Navy was called upon to play an

Fig. 3.9. Samuel Mitchell of HMS *Harrier* carrying Commander Hay at the Battle of
Gate *pā*, 29 April 1864

increasingly active role supporting British interests in the Indian Ocean
littoral. The spice monopoly held by the Dutch East India Company
(VOC) had been broken by Sir Stamford Raffles, who temporarily
extended free trade over Java, and founded Singapore in 1819. Alongside
Malacca and Penang, Britain controlled the key choke points in the
commercially valuable Malacca Strait connecting China and India, and, in
1843, Captain Henry Keppel of HMS *Dido* was appointed as Senior Naval
Officer in the Straits Settlements. On 17 March that year, he dined with
the 'White Rajah' James Brooke, a former Indian Army officer who had
been made Governor of Sarawak by the Sultan of Brunei after suppressing
a rebellion there. Keppel and Brooke shared a desire to end Malay piracy

and agreed that they must first eradicate the pirate strongholds in Borneo's interior. After a month-long campaign, a treaty was ratified by the Sultan on 28 May 1847, which ceded Labuan Island to the British under Brook's governorship.

In January 1838, the British extracted territorial concessions including Aden from the Sultan of Lahej, Muhsin bin Fadl, but encountered opposition from the Abdella tribe, who opened fire on the East India Company corvette *Coote*. The Commander in Chief of the East Indies Station, Rear Admiral Sir Frederick Lewis Maitland, dispatched an expedition led by HMS *Volage*'s Captain Henry Smith to seize the town in January 1839, and it became an important coaling station, virtually equidistant between Zanzibar, India and, from 1869, the Suez Canal. A month after Aden's annexation, Maitland commanded a naval force comprising HMS *Wellesley*, *Algerine* and Indian troops that captured the port of Kurrachee (Karachi). Four years later, the whole of the Sind was under British rule.

When the First Anglo-Burmese War broke out in 1824, Britain sought to extend its Indian frontier to secure Bengal against future Burmese aggression. In August the following year, the Admiralty pushed for a section of Arakan to be retained as a naval station, which led to the whole region being annexed by 1826.

Natural resources of strategic importance to the Admiralty became a catalyst for conflict. The navy valued teak since an English oak shortage during the Napoleonic Wars prompted the development of Bombay's dockyards. There the Wadia shipbuilders used Malabar teak to refit and construct vessels for the Royal Navy, including the 74-gun HMS *Wellesley*, and in Zoroastrian tradition they hammered an engraved silver nail into their ships' keels to ensure good fortune. By 1834 it was estimated that an 84-gun ship cost over £20,000 less to build in Bombay, with it 'universally admitted that a Bombay teak-built ship is 50 per cent superior to vessels built in England'. Colonial representations were carved into the figureheads and names of Royal Navy warships and then projected to the rest of the empire, such as the 84-gun HMS *Asia*, launched in Bombay in 1824.[28]

In Burma, British merchants collaborated with Martaban officials to secure teak from local competitors, and throughout the 1840s they called

Fig. 3.10. Teak figurehead of HMS *Asia*

for Britain to 'send a gunboat, show the flag' and annex Rangoon in support of free trade. It was the Palmerstonian belief that,

> as the Roman, in days of old, held himself free from indignity, when he could say *Civis Romanus sum*; so also a British subject, in whatever land he may be, shall feel confident that the watchful eye and the strong arm of England, will protect him against injustice and wrong.

Upon hearing reports of the arrest and abuse of British subjects, Commodore Robert Lambert volunteered to visit Rangoon in October 1851, where he attempted to extract compensation worth £1,000 from the local Governor. The arrival of Lambert's squadron at Rangoon was celebrated by the missionary Eugenio Kincaid, as:

> news too good to be true [ . . . ] the emotions of joy and gratitude diffused can only be understood by those who have seen and *felt* the deeds of insane and brutal tyranny – the navy guard spoke to us of *peace and security* [ . . . ] God's own hope, he will hear our prayers for deliverance.[29]

European traders and Christian missionaries viewed the navy as a means to further their respective interests, opening up new areas for commercial exploitation and religious conversion. Upon hearing from Kincaid that hundreds of British subjects had been tortured to death, and that the Governor had threatened to set Rangoon ablaze and behead any European greeting his squadron, Lambert refused to negotiate with 'a person whose conduct renders him totally unworthy of any respect' and 'unfit to be entrusted with the lives and property of British subjects'. He established a blockade, seized the king's royal barge, and silenced the Burmese gunfire with a broadside from his frigate, justifying his actions as necessary to uphold British prestige: 'Had I quitted Rangoon without some demonstration to show my determination to have our demands complied with these ignorant boasting people would have immediately declared it was from want of prowess'.

The Indian Governor-General Lord Dalhousie decried that 'these commodores are too combustible for negotiations', but following the rejection of Britain's ultimatum he saw 'no alternative but to extract reparation by force of arms'. It was hoped that a 'knockout blow' might allow for the resumption of informal British control, but the Second Burmese War ended with the annexation of Lower Burma on 20 January 1853.[30]

On the evening of 16 July 1857, HMS *Shannon* and *Pearl* were redeployed from Hong Kong and Singapore to deal with the rebellion which had erupted in India. While the mutiny of the Indian Army was sparked by the distribution of new Enfield rifles and cartridges believed to be greased with cow and pig fat, which contravened Hindu and Islamic custom, the uprising spread to civilian groups who opposed the extension of British East India Company rule and Governor-General Dalhousie's

Westernising reforms. Upon the ships' arrival at Calcutta in August, artillery was disembarked and Naval Brigades were formed totalling almost 800 men. In November, the *Shannon* played a key role in relieving the European residents of Lucknow, the capital city of Awadh, who had been besieged for six months. During the battle, the brigade's Captain William Peel, son of the free trade advocate and former Conservative Prime Minister Sir Robert Peel, 'behaved very much as if he had been laying the *Shannon* alongside an enemy frigate', with sailors and marines fighting at close quarters as Europeans were evacuated under the cover of naval rockets and gunfire. Four members of the brigade were awarded the Victoria Cross, including William Hall, the first black sailor and Nova Scotian to receive the medal, and Lieutenant Thomas Young. Together, they managed to breach the wall of the Shah Najif mosque with a 24pdr after the other gun crews had been killed or wounded at close range.

On Boxing Day the following month, HMS *Pearl*'s brigade fought alongside Sikh and Gurkha regiments, defeating 6,000 rebels at the battle of Soanpur. The victory was politically valuable, securing the support of the local Rajah, who had been threatened by the rebels, and restoring confidence in British rule between the Gogra river and Nepal. While a number of locals fetched water for the troops, keen to befriend the victors who paid for what they used, those villages which supported the uprising were burnt down after their inhabitants had an hour to pack and leave. Some sepoys were strapped to the mouth of field guns and blown away, while others were hanged. These scenes caused some naval personnel, such as *Shannon*'s Edmund Verney, to question this 'war in which the worst passions are likely to be excited, and without doubt dreadful scenes have been enacted [ . . . ] yet we claim to be the most enlightened nation in the world'. While Verney recognised that the professed benefits of British rule – the free market, Christianity, taxes – were considered a curse by some Indians in the subcontinent, he continued supporting the 'civilising mission' elsewhere in the empire when he commanded the gunboat HMS *Grappler* in British Columbia during 1862.

The Indian rebellion was declared over on New Year's Day in 1859 and the *Pearl*'s crew rejoined their ship on 2 February, having fought 26 engagements over 17 months ashore with just one man killed in action, though 18 died from other causes. The *Shannon*'s 68pdr guns would be left at Lucknow to guard its citadel for almost a century, and its brigade returned to Calcutta on 12 August, having lost 20 men to the enemy and 83 to disease, including Peel. There they walked beneath a triumphal arch inscribed with 'Welcome Hearts of Oak', before being greeted by girls throwing flowers, regimental bands, and 'one continued cheer from the street down to the river', followed by a civic reception at the Town Hall decorated with paintings of Britannia, Neptune and vanquished sepoys. Upon their return, the *Shannon's* crew cemented this imperial bond between the Royal Navy and India by donating £350 for a monument to be erected at Southsea.[31]

### From Cape to Cairo: Egypt and South Africa

Britain's interests in Africa increased with the opening of the Suez Canal in 1869, constructed with French money and local taxes raised by Egypt's Khedive ruler. Eight years later, Khedive Ismail was forced to sell his shares in the Suez Canal Company as he attempted to re-balance his government's deficit, which made Britain the controlling partner. Anglo-French influence over the country's ruler and its domestic affairs prompted an Egyptian Army mutiny in 1879, followed by a nationalist revolt two years later, which swept the military officer Ahmed Urabi to power in February 1882. Britain was worried that without the collaborative Khedive, Egypt might renege on its debts and nationalise the Canal, while reports were also received of rising tensions between locals and Europeans in Alexandria. Though Alexander Malet, the Minister of War, warned that a 'fleet is a menace likely to lead to disturbances and not to protection', the Cabinet decided it was better to cover itself in case British subjects needed evacuating, and on 20 May Admiral Sir Frederick Beauchamp Seymour arrived off the coast of Alexandria with HMS *Invincible*, a French battleship and four gunboats. This provoked the nationalists further, as anti-Christian demonstrations on 11 June turned into a massacre in which over 50 Europeans – including an engineer from HMS *Superb* – and 100 Egyptians were killed. This violence stoked the flames for imperial intervention in what became framed as a struggle of Christian civilisation versus Oriental barbarism. By now, Seymour had 15 ironclads and a force of gunboats under his command,

Fig. 3.11. Men of HMS *Emperor of India* at Cairo, *c.*1920

and after Alexandria's fort ignored his ultimatum to surrender on 10 July, the fleet opened fire on the city at 7:00 the following morning. By 17:30 that evening, they had silenced the defenders' guns and troops were landed. Captain Charles Beresford set up 'drumhead courts', where Egyptians found guilty of crimes against Europeans were tied to acacia trees in the city's main square and shot. Full occupation of the country was sanctioned by the British government on 18 July, justified on grounds of protecting the Suez Canal, which was seized by a Royal Navy amphibious assault on 20 August. Colonisation of Egypt helped trigger Europe's 'Scramble for Africa', as Britain's imperial rivals sought territorial compensation elsewhere, beginning with France in West Africa.[32]

In South Africa, the British settlement of Port Natal owed much of its early existence to the Royal Navy, since being founded by Lieutenant Francis Farewell in 1824. Its governor, Sir Benjamin D'Urban, publicly thanked Commander Edward Stanley of HMS *Wolf* after the ship defended the town during the Xhosa War of 1834–5. While, on 26 June 1842, the schooner HMS *Conch* and frigate HMS *Southampton*, led by Captain Thomas Ogle, helped break the Boer blockade of Durban, before Natal became a formal British colony the following year.

Britain retained a significant strategic interest in the region when the discovery of gold in the Transvaal produced an economic and political shift towards the Boer Republic. When this climaxed with the Boer War of 1899–1902,[vii] the Royal Navy logistically managed the movement of approximately 450,000 imperial and colonial troops to South Africa from Britain, India, Australia, New Zealand and Canada, while transporting thousands of Boer prisoners of war for incarceration in India and the colonies of St Helena, Bermuda and Ceylon. Its supremacy at sea limited the material assistance arriving from imperial rivals who sympathised with the Boers, such as Germany. Royal Navy warships monitored traffic entering Delegoa Bay in Portuguese Mozambique, and searched suspected merchant vessels for arms and ammunition which might be smuggled across the border. They occasionally exchanged fire with Boer patrols on the coast, as occurred with HMS *Partridge* in Saldanha Bay on 10 December 1901, while Naval Brigades amounting to 1,400 personnel and 60 guns fought ashore alongside local forces such as the Natal Naval Volunteers, who were formed in Durban on 30 April 1885 and numbered 123 men in October 1899.[33]

The most famous naval action of the campaign was the relief of Ladysmith by the bluejackets of HMS *Powerful*. On 12 October 1899, 20,000 Boers – twice the number of British defenders – crossed the Natal border. On 25 October, a telegram was received from General Sir George White at Ladysmith that, 'in view of heavy guns brought by General Joubert from the north, I would suggest the Navy be consulted with the view of their sending here detachments of bluejackets with guns firing heavy projectiles at long ranges'. Two days later, HMS *Powerful* was sent from Simon's Town to Durban, where the ships' 4.7-inch and 12pdr guns were unloaded with a Naval Brigade to march them to the besieged town. Their arrival altered the course of the battle for Lombard's Kop, giving the defenders weapons that could finally match the range of the Boers' 6-inch Creusot guns. This made the attackers pause and maintain their distance, bringing an end to the siege by February, as one Boer recalled:

It is very dangerous to attack the town. Near the town are two naval guns from which we receive very heavy fire which we cannot stand. I think there will be

---

[vii] See also Duncan Redford and Philip D. Grove, *The Royal Navy: A History Since 1900*.

Fig. 3.12. Naval Brigade from HMS *Terrible* at Durban *c.*1899

much blood spilt before they surrender, as Mr Englishman and his damn sailors fight hard.

The exploits of *Powerful*'s bluejackets earned them a hero's welcome when they returned home in May 1900. Hundreds of spectators broke through barriers to cheer the ship's arrival at Portsmouth, where streets were bedecked in red, white and blue, patriotic colours which the Town Hall's electric lights formed into the letters 'RN'; children from the Royal Seamen's and Marines Orphan's Home serenaded the crew with 'The Lads of the Naval Brigade' and 'Home Sweet Home', while boys in sailor costumes fired a salute from a model 4.7-inch gun. Britain eagerly expressed its gratitude to the *Powerfuls* for their service to the nation, Natal and the empire. This came from village committees presenting watches to naval residents, to thousands of Londoners lining the crew's parade through the capital, where imperial financiers Lloyd's dined them, and even Queen Victoria reviewed the brigade personally at Windsor Castle.[34]

To capitalise upon this fervour, the Royal Navy's Portsmouth recruitment office displayed *Powerful*'s white ensign, which had flown in Ladysmith during the siege. Navy League member and *Daily Telegraph* journalist Archibald Hurd was also inspired by the Naval Brigade's exploits, having been asked to publicise the service with other navalist

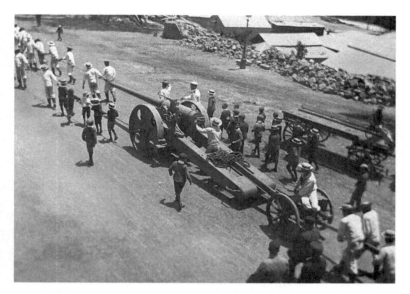

Fig. 3.13. Sailors and gun from HMS *Doris* ashore during the Boer War

writers including Rudyard Kipling. Published in the Boer War's final year, Hurd's *How Our Navy is Run* emphasised the national and imperial importance of the Senior Service, 'who guard our shores, protect our very daily bread as it is borne over the ocean from far distant lands, and safeguard the overseas empire'. Here, the 'Jack Tars' of Ladysmith highlighted the Royal Navy's vital role in Britain's imperial 'civilising mission',

> as a fighter Jack upheld his character so splendidly during the siege of Ladysmith, and through the earlier part of the South African campaign; as in China, Egypt, Ashanti and elsewhere [ . . . ] the beleaguered Natal town owed its salvation to the Naval Brigade and naval guns which had proceeded across country from Durban [ . . . ] naval detachments have many times, during the past ten years or so; distinguished themselves in land fighting [ . . . ] He seems to have a happy knack of adapting himself to varied circumstances in perhaps a greater degree than his comrade in the Army and it is for this reason that he is so often selected for the work of chastising a coast tribe or the chief of the interior of the Dark Continent.[35]

## Oil imperialism

The second industrial revolution brought developments in oil refinement by the end of the nineteenth century, which offered significant advantages over coal as a fuel source for the Royal Navy's warships. Future First Lord of the Admiralty, Winston Churchill, recognised that the physical 'ordeal of coaling ship exhausted the whole ship's company. In wartime it robbed them of their brief period of rest; it subjected everyone to extreme discomfort'. It was impossible to refuel at sea, meaning a quarter of the fleet was always in port, while coal produced smoke which provided a target for the enemy. Admiral John 'Jackie' Fisher, First Sea Lord from 1904, was also a self-confessed 'oil maniac', attracted by the ability to replenish at sea and the savings in weight and manpower that oil offered in requiring half the number of stokers. It also brought greater speeds that could only otherwise be achieved by lengthening battleship designs to accommodate larger boilers, making them too big for existing docks, which would need rebuilding at extreme cost.[36]

This raised the issue of supply, for while Britain was self-sufficient in coal, it possessed few oil reserves in its empire; India and Canada produced just 1.2 per cent of global output in 1900. In July 1903, Burmah Oil, which supplied lighting to Indian peasants, was approached by the Admiralty's Director of Contracts about whether it could provide oil 'for use as liquid fuel'. With the United States, Rumania and the Dutch East Indies producing the majority of naval fuel oil, the Admiralty argued in June 1904 that India's oilfields were an imperial interest, it being strategically vital that the Royal Navy's supply 'should come from within the Empire, should be available in war as well as in peace and should not be under the control of any foreign Trust or syndicate'. The following month it pushed the Colonial and India Offices 'to reserve the [oil] concessions for the exploitation by British money and in the interests of the Colony and of the Navy'. The oil committee formed by the Admiralty and Colonial Office drew up a prospecting licence for the whole empire, which stipulated that all concessionaries must be British, crude oil must be refined within British territory, and the Admiralty reserved the right of 'pre-emption' during wartime. British companies like Burmah Oil used Admiralty support to further their own business interests, thwarting rivals such as the Shell-Asiatic Alliance from establishing production bases in Burma by highlighting their 'foreign' connections. In 1905, the Admiralty persuaded Burmah Oil to invest in the small British oil company in Persia, established by William D'Arcy four years earlier. This became the Anglo-Persian Oil Company in April 1909, and strengthened Britain's position in that strategically important, oil-rich region against old and new imperial rivals, Russia and Germany.[37]

Anglo-Persian pushed the British government to award it a large forward contract to supply oil for both the Royal Navy and the Indian Railway, and requested financial and diplomatic assistance from the Admiralty, India and Foreign Offices to expand production and obstruct foreign competitors from acquiring concessions in Mesopotamia. To ensure that the navy had access to increased quantities of oil during war, the Indian Railway was required to purchase the peacetime excess. Fisher personally visited the India Office to argue Anglo-Persian's case and convince the Indian government to accept the costs of converting its train engines to oil. When Churchill became First Lord of the Admiralty in October 1911, one of his first acts was to form a committee under the Fourth Sea Lord, Captain Pakenham, to consider the issue of oil fuel. This led to the decision in April 1912 to lay down the *Queen Elizabeth*-class, the first battleships fired solely by oil, giving them the speed to outmanoeuvre the coal-fired Dreadnoughts of the German Navy. For Churchill, this order produced a domino effect on the rest of the fleet:

> The fateful plunge was taken when it was decided to create the fast division. Then, for the first time, the supreme ships of the navy, on which our life depended, were fed by oil and could only be fed by oil. The decision to drive the smaller craft by oil followed naturally upon this. The camel once swallowed, the gnats went down easily enough.[38]

Four months later, Fisher admitted that they were 'all bitten' with 'internal combustion rabies'. On 11 July 1913, the British Cabinet agreed that 'in view of the vital necessity to the Navy of a continuous and independent supply of oil in the future it was desirable that the Government should acquire a controlling interest in trustworthy sources of supply'. Churchill cemented this with a speech six days later, in which he reasserted the Royal Navy's oil policy:

The Admiralty should become the independent owner and producer of its own
supplies of liquid fuel, first, by building up an oil reserve sufficient to make us
safe in war and able to over-ride price fluctuations in peace; second, by
acquiring the power to deal in crude oils as they come cheaply in the market
[ . . . ] third [ . . . ] we must become the owners, or at any rate, the controllers
at the source of at least a proportion of the supply of natural oil which we
require.

To this end, the Admiralty sent a commission under Rear Admiral Sir
Edmond Slade to investigate Anglo-Persian's resources. In June 1914, the
government acquired a controlling 51 per cent stake in the company,
placing two directors on its board with the power of veto in matters of
foreign and naval policy, and securing a contract to supply oil to the Royal
Navy for 20 years. Just weeks later, the British Empire would be at war.[39]

CHAPTER 4

# Imperial and Colonial Culture

The impact of Britain's imperial endeavours was not only felt overseas and within the political and economic spheres; it also resonated culturally across broader society at home. The exploits of empire adventurers inspired novels, poetry, art and music, attracting a popular audience for whom Royal Navy officers served as archetypal heroes for an imperial culture they would help spread.

Britain's imperial endeavours overseas were ideologically supported by popular culture at home. In November 1739, after Vice Admiral Edward Vernon captured the Spanish settlement of Porto Bello in Panama, James Thomson penned his most famous of patriotic anthems, 'Rule Britannia'. Its bombastic timbre coveys the boundless energy of British spirit, which dictates that 'Britons never will be slaves' because they 'rule the waves'. Echoing the arguments of Tudor courtiers like John Dee, Britain's right to command others is portrayed as an act of providence, for 'at Heaven's command' the island nation was 'blest' by its geography which 'arose from out the azure main'. Its greatness and freedom rested on this maritime strength, protected by the 'native oak' and 'manly hearts' of the Royal Navy, who helped enrich Britain by spreading its concept of liberty across other continents, so that its 'cities shall with commerce shine [ . . . ] And every shore it circles thine'.

The mutiny of HMS *Bounty* inspired George Keate to support Bligh by publishing a set of verses for him in 1793, where he is depicted as the successor to his former Captain, and the rightful inheritor of James Cook's imperial mission. Blessed by the 'hand Divine' of 'Providence',

which guided him and those cast adrift 'thro' seas untry'd', Bligh is urged
to redeem himself by spreading Britain's influence and discovering new
territories:

> what Cook was, hereafter Bligh may be.
> Where cannot Britain's dauntless sails extend?
> Go search our tracts, and nations yet unknown,
> 'Midst her proud triumphs some fresh laurels blend,
> And with thy country's fame augment thine own.[1]

Nelson's victory at the Battle of the Nile in August 1798 had a huge
cultural impact back in Britain, as Egyptian-style clothing became the
fashion of the day, accompanied by jingoistic prose. Poet Laureate Henry
James Pye's *Naucratia, or Naval Dominion. A Poem* was published that year,
dedicated to the Royal Navy which had extended Britain's 'discoveries,
increased her commerce, and established her dominion beyond the
example of former ages'. For Pye and his public, the connections between
naval power, prosperity and empire were indisputably intertwined,
meaning 'Britain's conquering flag had oft unfurl'd, to every region of the
peopled world'. Like Dee and Thomson, he ascribes this national
ascendance to the country's natural qualities which produced 'the rise and
progress of an art' of seamanship, 'which has ever been the peculiar glory
and strength of Great Britain'. Though the *Critical Review* acknowledged
the poem's popularity, it challenged Pye's imperial sentiments and his
description of the Union Jack as 'The holy labarum of liberty', for 'alas!
The flag of England is not considered as the ensign of liberty in the east or
in the west, or on the shores of Africa', or in America, where the War of
Independence was not long passed. The year after *Naucratia*, William Lisle
Bowles published *Song of the Battle of the Nile*; like Keate, he used the
defeat of Catholic France to demonstrate divine approval for Britain's
Protestant imperialism.[2]

The London 'establishment' were shook by a scandal in 1808 involving
the Duke of York, King George III's son and Commander in Chief of the
Army. Allegedly his mistress, Mrs Clarke, had been using cash and sexual
favours to bribe a number of ambitious officers who sought the Duke's
patronage and rapid promotion. It was believed that this corruption of the
army compromised Britain's war effort against Napoleon and threatened

to produce a political and social revolution similar to France's. The navy provided a bulwark against this by representing purer ideals of national character, such as heroism and self-sacrifice, which Nelson had epitomised. Further royal sex scandals, such as the 1820 Queen Caroline affair, aroused political concerns that British colonialism, which materially profited from the exploitation and trade of exotic wares, had caused corruption to spread like a venereal disease from the East to the British Isles. Debauchery became a foreign import sullying the sanctity of the British race and meant that to purify itself, Britain had to embark on a moral crusade to eradicate the source of the problem by spreading its more positive characteristics overseas.

The taint on Britain's aristocracy saw the middle classes inherit their ideals of chivalry, aided by the literary representation of naval heroes following the cult of Nelson. The writings of Robert Southey, Samuel Taylor Coleridge and Jane Austen, whose brothers served in the Senior Service, portrayed naval officers as romantic, chivalrous, paternalistic and commanding gentlemen, resistant to political and moral corruption. They contributed to the spread of a gendered view of the Royal Navy and its imperial acts which redeemed British manliness from the feminised ruling classes and their sensual wantonness.

In 1809 and the aftermath of the Clarke scandal, Admiral Sir Alexander Ball died. He had been Commissioner of Malta since the island was occupied in 1799 to serve as a naval base, and Coleridge was his secretary from 1804. In *The Friend*, Coleridge eulogises the man who was called 'father' by both his naval subordinates and Maltese civilians; Ball regularly held open meetings with the colonial public to address their concerns, an individual act of paternalism which Coleridge saw as symbolic of Britain's greater imperial mission,

> there are I trust many who will think it not the feeblest reason for rejoicing in our possession of Malta, and not the least worthy motive for wishing its retention, that one source of human misery and corruption had been dried up. Such persons will hear the name of Sir Alexander Ball with additional reverence as one who had made the protection of Great Britain a double blessing to the Maltese, and broken the 'bonds of iniquity' as well as unlocked the fetters of political oppression.

By banning Malta's seraglios, Ball elevated himself above the Duke of York's own penchant for such indecent fraternising, while presenting British rule as more moral and benevolent than the Catholic orders that preceded it. He was thus depicted by Coleridge in chivalric terms, akin to 'an English knight instituting a new colonialism that protected but did not take advantage of a people too weak to defend or govern itself'.[3]

Thirteen editions of Southey's biography – the *Life of Nelson* – were published between 1813 and 1853 for a voracious British public, and they presented the parson's son as a dutiful, patriotic and courageous gentleman who had risen through the ranks on merit, making him a better leader than their 'high-born' rulers in London. He also represented a grander form of patriotism, protecting not just Britain's shores, but also the source of its greatness, by thwarting France's attempts to steal its colonial possessions; in Egypt, Aboukir in India, and St Vincent in the Caribbean, Nelson halted Napoleon's forces,[viii] while his resolution to 'discharge [ . . . ] my duty' towards the Navigation Act saw him risk civilian trial by seizing four American vessels involved in an illegal trade with British planters.

The example of Nelson was even more telling because of his relationship with Lady Hamilton. She embraced the Oriental fashion of the time, wearing Indian shawls to dress-up provocatively in the style of Circe and Cleopatra, leading satirists such as James Gillray to depict her as an Oriental temptress, and the bare-breasted Cleopatra to Nelson's Mark Antony. While the Orient's sexual intrigue had seemingly seduced Nelson like an East India Company nabob, Southey emphasised how the Admiral then saved himself by escaping into the masculine, Protestant culture of the navy to 'do his duty'. He was therefore an exemplar of paternalistic, manly virtue, both for Britain's imperial mission and those who would venture into the empire, possessing the self-discipline to resist corruption by the exotic and thus fit to command others. For Southey, conflict cleansed character, by conquering and governing imperial territory, young British gentlemen were readied for leadership back home. The *Life of Nelson* influenced Victorian society from the highest echelons down. The young Princess Victoria expressed her enjoyment of it to Southey

---

[viii]  See also Martin Robson, *A History of the Royal Navy: The Napoleonic Wars*.

*A* COGNOCENTI *contemplating y̆ Beauties of y̆ Antique*.

Fig. 4.1. James Gillray's cartoon of Sir William Hamilton inspecting his antiquities, all of which refer to his wife and Lord Nelson who are pictured as 'Cleopatra' and 'Mark Antony'

personally in 1830; Sir Robert Peel recommended the book to his son William, who would serve in the navy before he died suppressing the Indian rebellion; and it became an established school text by the end of the century.

Captain Frederick Marryat had joined the Royal Navy in 1806 at the age of 14, becoming a midshipman under Lord Thomas Cochrane aboard the frigate HMS *Imperieuse*, before serving in the First Anglo-Burmese War. A keen writer, in 1822 Marryat combined his interests to argue for the navy's professionalisation in his pamphlet, *Suggestions for the Abolition of the Present System of Impressment, in the Naval Service*, which he hoped would restore 'that ardent and enterprising character which exalted the English navy to the zenith of its splendour'. Seven years later, whilst commanding HMS *Ariadne*, Marryat penned and published his first novel, *The Naval Officer, or Scenes in the Life and Adventures of Frank Mildmay*. Drawing on his firsthand experience of the service, the novel's realism made it a huge literary and financial success; the character of Mildmay was clearly modelled on Marryat's notorious former commanding officer, Cochrane, who had been discharged from the navy after speaking out against corruption in parliament. Yet he represented 'the very flower of chivalry', and Marryat's figure of Mildmay helped remind Britain of Cochrane's heroism before he was reinstated to the navy in 1832. Upon his death in 1860, *Punch* declared that Cochrane had been betrayed by 'the crawling worms that in corruption breed'.[4]

Having proven the public thirst for tales of naval adventure, Marryat followed upon this success in 1830 with the publication of *The King's Own*, drawing the praise of the American author Washington Irving who thought 'the chivalry of the ocean quite a new region of fiction and romance, and to my taste one of the most captivating that could be explored'. Readers of *The King's Own* witness India through the eyes of Captain M. and his crew, who provide a cultural reference point they recognise and empathise with when confronted with scenes depicting Indian otherness and British moral superiority. The crew are repulsed by the 'sickening' practice of children being sacrificed by their parents and cast into the holy river to appease the gods, a scene which secures their support for British intervention to suppress the 'savage' custom. The naval officers possess a self-discipline which is resistant to the Orient's corruption, that the locals lack, allowing them to guide Indians towards British standards of 'civilisation', while convincing the reader of the necessity of Britain's imperial mission there and fortifying their own sense of national character.

More than entertainment, the naval genre provided a moral compass at a time when readers in Britain and across the empire were questioning the imperial mission, leading Robert Louis Stevenson to comment that 'stories of our sea-captains [ . . . ] full of bracing moral influence, are more valuable to England than any material benefit in all the books of political economy'. Following the tradition laid down by Austen, Southey and Coleridge, Marryat's naval officers were depicted as the virtuous embodiment of patriotism, self-reliance, courage, paternalism and duty, which made them better leaders and gentlemen. They helped form a myth of national character and manifest destiny, which Victorian imperialists used to justify extending British rule overseas so that others could be 'civilised' by imbibing these qualities. As late as the 1920s, children were taught the importance of the Royal Navy to Britain's world role in stories such as Rupert Chesterton's *The Captain of the Phantom*, where the fictional HMS *Phantom* is sent to restore liberty, law and order to a South American state which has succumbed to banditry and revolution, and is greeted by 'throngs of people that crowded every point of vantage to cheer' their emancipators.[5]

The Russian war scare of 1877–8 gave birth to a song by G.H. MacDermott and G.W. Hunt, which included the chorus line 'we don't want to fight but by jingo if we do, we've got the ships, we've got the men, we've got the money too'. While 'jingo' was used as a minced oath for 'Jesus', the song's popularity in music halls and public houses saw the word fall into general usage to represent both an especially vocal and aggressive form of patriotism and the genre of songs it inspired. At the height of this fervour, W.S. Gilbert and Arthur Sullivan penned *HMS Pinafore; or, The Lass That Loved a Sailor*, which opened in London on 25 May 1878 and ran for 571 performances, the second-longest running piece of musical theatre at that time. While the patriotic heroism encapsulated by the words and music were meant to satirise over-the-top nationalism, this was lost on many members of the audience, who believed that songs such as 'He is an Englishman!' were celebrations of national pride and glory. The song thus became nicknamed 'the jingo jingle' by Sullivan's friend and fellow composer Francois Cellier, and for a time rivalled the National Anthem in popularity. The success of the genre saw jingoistic songs take inspiration from the Royal Navy's real-life exploits, such as its bombardment of Alexandria in 1882 upon which T.S. Lonsdale's 'The Gallant Blue Jackets of England' is based. Though a supporter of closer imperial links between the white settler colonies of 'Greater Britain', the Liberal Minister Charles Dilke declared that jingoism made full-scale intervention in Egypt inevitable, as the Royal Navy's 'bombardment of Alexandria like butchery is popular'.[6]

Beyond Britain, the Royal Navy was a crucial cultural adhesive for binding the empire's young settler societies together with the mother country. Wherever a British naval base or warship was present, the service occupied a prominent position in the social life of the colony and in disseminating British imperial culture. Naval personnel at Victoria put on and performed in amateur theatre productions, including one in December 1868 featuring Sub-Lieutenant Sydney Smith Haldemand Dickens, the youngest son of famed writer Charles, whose name became attached to Dickens Point in British Columbia's Portland Canal. Dancing also featured prominently in colonial entertainment, helping to preserve the manners and customs of the old country, and was accompanied by Royal Navy bands and hosted upon the decks of its ships whenever one visited port. Naval dances were also important for the colonies' economic development, helping to raise funds for the construction of facilities such as hospitals. In the words of the American poet Edgar Fawcett, who grew up on Vancouver Island during the 1860s: 'The navy was the mainspring of Victoria in more ways than one. They took part in all public functions, furnishing music, help and flags, and by their presence in uniform brightened up and lent grace to the affair'. This intimate civil–naval relationship inevitably produced marriages between Royal Navy personnel and settler women, while others moved their families from Britain to live and retire in the colonies.[7]

The Royal Navy also influenced the development of indigenous cultural performances outside of European cliques. During the 1890s, *beni ngoma* emerged in towns along the Swahili coast of Kenya, including its major port of Mombasa, and was a dance performance that imitated the movements and rhythms of naval drills and brass bands that locals picked up by watching the Royal Navy (its name merges the English word 'band' with the Swahili for 'dance'). This grew into a competition between rival groups of young Swahili Muslims, which spread to the Tanganyikan coast by the mid-1910s. *Lelemama* was another East African dance form, and one that was traditionally the preserve of female elders who used it as a way of reinforcing their freeborn social status by banning slaves from performing it. Congolese slaves appropriated *lelemama*, spreading their own form via Zanzibar so that it became a female equivalent of male brass bands and an inclusive and integral element in Mombasan Swahili culture from the 1890s to the late 1940s. *Lelemama* associations also displayed Britain's colonial and naval influence, most notably in costumes modelled on Royal Navy uniforms, and in the titles their members adopted. The leader was called 'Queeni', while other performers would adopt names they associated with colonial power and prestige, particularly 'Admiral'. Apart from the Queen, these titles were traditionally held by men, thus *lelemama* allowed Swahili Muslim women to transcend social positions they were otherwise prohibited from.[8]

*Taarab*, a Swahili musical form famous for its distinctive poetry and sound, was the most popular leisure activity for urban Zanzibari women between World War II and the 1960s. One of its earliest and most prominent bands called themselves 'Royal Navy', inspired by the visiting ships and their crews. *Taarab* was about more than performance, it was a contest, and Royal Navy's main competitors took their name from a rival branch of Britain's armed forces, 'Royal Air Force'. The shows staged between these two bands 'drew [large] crowds [ . . . ] became the talk of the town', and were inspired by the service traditions of their namesakes. In one instance, Royal Air Force hired a plane to drop leaflets over the town, prompting Royal Navy's leader, 'The Admiral', to approach a naval captain in port to take the group out in his vessel:

> We had to show them up. So I rented a man-of-war and we went for a ride. We went out to the island, spent the day, had a picnic, and in the evening we returned to town. These were our games.

This generated huge excitement in the town, 'as fans watched, weighed the force and impact of each club's latest moves, and cheered them on'.

As with *lelemama*, an important part of the bands' identities and status resided in their attire, and 'the style and "class" of the uniforms' attracted women to join Royal Navy, just like some male motivations for joining the Senior Service. The uniforms inspired songs too, which Royal Navy used to mock Royal Air Force for copying their motif: 'They are indeed our uniforms, nor are there any others/if people imitate what is said they imitate words of meaning'. After the 1964 Zanzibar Revolution, the new socialist regime forced Royal Navy to change its name to *Wanabaharia*, or 'Those Who Work at Sea'.[9]

One of Zanzibar's first displays of cinematography was also due to the Royal Navy. During HMS *Vengeance*'s visit to the islands in May 1916, at the height of World War I, its flag officer, Rear Admiral Charlton, sought to raise contributions for the Zanzibar War Society by organising an evening of film screenings at the Victoria Gardens, across from the British Residency. The ship's band performed during the intervals, and the event drew a 'large audience', making it an 'unqualified success'.[10]

In Trinidad during World War II, West Indian ratings serving in the local Royal Naval Volunteer Reserve (RNVR) also used local musical forms to express their imperial patriotism. Just as sea shanties have been synonymous with the life of sailors for centuries, songs became an integral part of naval life for Trinidadians. Here they manifested as calypso, a form indigenous to and deeply engrained in the fabric of Trinidadian society, where calypsonians took on the role of social and political commentators. Sailors in the Trinidad RNVR carried this torch, the chorus of one such example being:

> Commander Lindsay say
> Cheer boys cheer
> With unity and the TNV [Trinidad Naval Volunteers]
> We gonna conquer Germany.[11]

Naval songs for Trinidadians were more than ditties sung to relieve boredom and pass the time, they were expressions of national identity. Here though, traditional cultural forms were being used not as anti-colonial nationalistic expressions, but to reaffirm Trinidadian identity with the imperial cause, united in their mutual struggle against Germany.

There was clearly a degree of loyalty to Britain amongst those Trinidadians who served in the Trinidad RNVR, and who chose to vocalise their 'unity' in this manner. The above chorus was in fact a clear pastiche of a famous patriotic calypso song from World War I, written by Henry Forbes:

> Run you run, Kaiser William, run your run, *repeat* [sic]
> Hear what Kitchener say: cheer, boys, cheer
> With surety and sincerity
> We going conquer Germany.[12]

Another naval-themed calypso from World War II was written by 'Atilla the Hun'; it celebrated the destruction of the German 'pocket battleship' *Graf Spee*, which had sunk nine merchantmen in the Atlantic, affecting food supplies to Trinidad:

> The sinking of the Admiral Graf Spee
> Must remain incontestably
> A monumental testimony
> To Britain's naval supremacy.

Such songs tied into calypso's rooted associations with battles and warrior deeds. The origins of modern calypso can be traced in nineteenth-century calinda chants, which accompanied stickfighter duels, and further back to the *djeli*, West African tribal singers who praised the heroic exploits of the tribe's warriors and reviled its enemies. Calypso was overwhelmingly a male form of cultural expression, and the connection between masculinity and warriorhood – a prominent feature in the formation of Trinidadian identity after emancipation – continued into the twentieth century. When Trinidadians enlisted in the Trinidad RNVR, they were doing so for more than financial betterment, imperial patriotism and national honour; they were responding to the calypso warrior's call. Wartime naval service allowed Trinidadians to prove themselves as men, and, when they sang calypsos, they framed their own heroic deeds as part of this longer warrior tradition and reaffirmed their own masculinity.[13]

## Sport

In 1912, the artist Sigsmund Goetze was commissioned to produce a mural series to decorate the Foreign Office and depict the mythical life of Britannia. Beginning with her conquest by ancient seafaring tribes, it shows Britannia mature into the warrior *Bellatrix*, equipping her sons for their expansion of the 'island race' and the pacification of the world through the gift of physical exercise and sport. It was no coincidence that the codification of organised sport developed at the same time as the British Empire was reaching its height. Like naval fiction, theatre and music, sport provided more than popular entertainment. It also helped inculcate and hone those attributes that young British men would need to perform their naval and colonial duties – duty, discipline, manliness, teamwork, self-sacrifice, fitness, fortitude and honour – to ensure that Britain not only survived but prospered in the Darwinian struggle for imperial dominance. The Royal Navy's technological transition from sail to steam also meant that more formalised exercise was needed to maintain the physical condition of the men and project a positive image of the nation's health to both its rivals and the ruled. This was developed by introducing gymnastics aboard ships, and arranging competitive matches against representative teams ashore. Organised sports became an essential part of naval life and training, and the Royal Navy was a major mechanism in their spread overseas, so that British games became the dominant forms internationally, and meaning others had to play by their rules and accept their cultural values.[14]

Competitive sport established footholds in colonies which hosted a Royal Navy base. Cricket, football, rowing, shooting, athletics and horse racing competitions were all actively promoted in British Columbia by Royal Navy personnel stationed at Victoria Fort from 1843 and Esquimalt from 1865. As in their literary representation, naval officers represented paragons of British behaviour and tradition for settler societies where they were also prominent members. The wealth generated by gold rushes created an inverted social pyramid in the colony, which blurred the old class boundaries of Britain and meant that the more common people attempted to elevate their position by mimicking the affectations of polite behaviour. Royal Navy officers lent their leadership, organisation and expertise in cultivating the colony's social cohesion by instituting local events and entertainment for its settlers. They also served as role models for aspirational colonists looking to acquire respectability through shared sporting endeavours. The first naval regatta in Victoria reportedly took place in June 1859, involving HMS *Pylades, Tribune* and *Satellite*. While such events served a practical purpose in training naval personnel, they quickly became highlights in the social calendar that brought colonists together for entertainment:

> A large number of spectators were present, among whom were many ladies, the varied colours of whose costumes imparted a lively gaiety to the same. During the first day's race two punts manned by the coal heavers of the ships, in grotesque costumes, who used their shovels for paddles, performed some novel and amusing feats.[15]

When they were not 'pacifying the natives', the gunboats *Forward* and *Grappler* and cutters from HMS *Devastation* and *Mutinie* contested three-mile races for a $100 prize. Visiting warships also participated, and Queen Victoria's birthday in 1870 brought out 5,000 spectators to watch ten Royal Navy crews from HMS *Zealous, Liverpool, Lifley, Picone, Pearl, Seylia, Charybdis, Endymion, Sparrowhawk* and *Boxer*, contest a regatta reported as 'the largest ever held on the American side of the Pacific'. With the ships' bands urging them on with renditions of 'Rule Britannia' and 'British Grenadiers', the crews competed in double sculls, four-oared shells, canoes, copper punt races and a 'duck hunt', whereby a man in a flat-bottomed skiff had to elude four men in a gig. Such exhibitions inspired the civilian youth to take to the water themselves, with the navy's Captain Nagle starting a rowing race in 1861 involving boys from the Collegiate School, while yachting also became a competitive recreation in the colony. This was quite a sea change for a settler community of miners, merchants and gentry who had not grown up on the water, and thus the navy helped reconcile them with their new maritime home.

Cricket competitions were also regularly held between colonists and the Senior Service, with matches played against visiting warships and between officers and ratings. The Royal Navy provided the material facilities needed for the sport's development, flattening and draining a section of Puget Sound Company land to create a pitch. This was also used for track and field competitions, including one held in July 1868 which attracted almost 2,000 spectators and prompted the *British Colonist* newspaper to declare that:

> The public of Victoria and vicinity surely fall under much obligation to Admiral and Mrs. Hastings, the Captains and Officers of the squadron on this station, for the day's superior amusement offered them [ . . . ] yesterday showed that distance takes aught from the success of similar entertainment in the colonies.[16]

Rugby matches between the Royal Navy, army and civilian teams were played on Vancouver Island from 1876, while rifling competitions naturally evolved from naval and military exercises. Horse racing was established with a two-day Spring Meeting held in May 1861, which drew over 2,000 spectators, and its success was followed by an October Race Meeting which featured a Stewards' Plate and Naval Stakes. Royal Navy officers even played the role of jockey, with Commander Robson riding 'Grey Arrow' in the Naval Purse. Robson was also a horse owner, and a founding member and promoter of the Jockey Club; he earned local fame and a $500 wager when he rode 'Butcher Boy' over three miles in six minutes and 35 seconds. Royal Navy gunboats were sent on more peaceful 'civilising missions', spreading Britain's sporting culture to people beyond its jurisdiction, as the *San Francisco Bulletin* reported on 1 July 1871:

> Whenever the British war steamer *Zealous* is in this port, there is a revived interest in the truly noble healthful game of cricket. It is infinitely more agreeable to have her send ashore a few cricket balls than it would be to receive any of her cannon balls.[17]

The cultural bonds established between colony and mother country through the navy's sporting missionary work were replicated elsewhere in the empire, formalised through clubs and associations which propagated imperial conceptions of control, class and status. Hunting was a pursuit particularly imbued with these principles, and Gibraltar's Royal Calpe Hunt was central to the formation of a colonial 'nobility' on the Rock. Hunting was believed to foster manliness, courage, hardiness, temperance, coolness and clear-headedness, and develop an *esprit de corps* amongst the British officer class serving far from home. Rear Admiral the Honourable C. Elphinstone Fleeming served as first master of the hunt, and naval officers participated having used their ships to ferry dogs from Britain and horses from North Africa. Members of the British royal family hunted in Gibraltar from the mid-nineteenth century, giving the Calpe Hunt its rare royal title and increasing its prestige for ambitious officers and colonists who sought elite status through this association. Hunting's appeal also lay in the regular social events that accompanied it, including balls and picnics.

A rarefied atmosphere was also cultivated by the formation of the Royal Gibraltar Yacht Club in 1829, the eighth oldest in the world and the first established outside the British Isles. To gain membership, one had to sail, own a boat and be socially acceptable, meaning that, in 1896, only 22 of its 70 members were civilians, just 6 of whom originated from the colony. In contrast, the club displayed crests from hundreds of Royal Navy vessels it hosted, and the position of commodore remained the preserve of British naval flag officers until 1941. It was the venue for receiving foreign luminaries, politicians and royal guests, and its awards night, held in the Governor's personal residence, was the social event of the year. Colonial yacht clubs again gave their socialite members entry into an elite world of cocktail parties, dances and dinners, which were as much, if not more, part of their appeal as the physical exercise of sailing.

One reason for the exclusivity of Gibraltarian society compared with British Columbia lies in its role as an imperial fortress of vastly greater strategic significance at the gateway to the Mediterranean than the Royal Navy's distant base at Esquimalt. While the Gibraltar Football Association was formed by civilians at the end of the nineteenth century, naval personnel had been involved in spreading the game and participated in their own local league, with each battleship possessing a unique team strip and attracting ships' companies of spectators. The navy and army's cultural and political predominance on the Rock meant that a civilian colonial identity did not challenge it until the first half of the twentieth century. Participation in sport required access to facilities, and these required land, which was at a premium on the Rock compared with the larger settler colonies and was primarily allocated for Naval and military use. The class and gender bar this created meant that ordinary

Gibraltarians resented the imperial armed forces and the wealthy local elite for restricting their access to recreational spaces and organised sports. This issue erupted following World War I in the campaign to form an elected municipal council, and it divided the Gibraltarian working class from 'the English' military they saw as colonial rulers. Land was handed over as part of a policy of Anglicisation, but while crossover matches occasionally took place, separate civilian and service football leagues continued after World War II. Consequently, a strong Gibraltarian identity came to be expressed in locally rooted team names such as Europa Club and Saint Bernard's, in contrast to 'imperial' rivals such as the Prince of Wales Club, Britannia FC and English United. As the Gibraltarian heavyweight boxer John 'the White Bomber' Ochello explained in reflecting upon his 1950s popularity: 'people were looking for something that was truly Gibraltarian – not English, or Military or Royal Navy [ . . . ] we needed something or someone to be proud of'.[18]

In Hong Kong, rugby matches between the armed forces, the Hong Kong Football Club (HKFC) and Hong Kong and Shanghai Banking Corporation served as a social hub for European colonists, like elsewhere in the empire. The game expanded as periods of economic and political upheaval affected the Chinese mainland, creating the Royal Navy base HMS *Tamar* in 1897, while the China Station acquired a summer anchorage at Weihaiwei to check Russia's lease of Port Arthur. For incoming British servicemen, rugby offered entertainment, fraternity and a cultural connection to home. While China dealt with the death of its Guangxu Emperor in 1908, precipitating the collapse of the Qing Dynasty in 1912, the accompanying military and sporting growth in Hong Kong saw the formation of a Triangular rugby competition in 1910. Contested by the Royal Navy, the army garrison and the HKFC, it became the colony's premier rugby tournament, and one which the navy dominated by winning 12 titles over the 1920s and 1930s. A combined Royal Navy and HKFC team also hosted matches against the touring Australia Universities in 1934 and New Zealand's Junior All Blacks two years later. The Royal Navy's encouragement of colonial sports clubs was not entirely altruistic. Organised bodies of healthy men possessing the right character could provide a source of ready-made recruits in times of crisis. It was hoped that the Royal Hong Kong Yacht Club in particular would provide

Fig. 4.2. Hong Kong harbour during the 1900s

'enthusiasts with the advantage of a knowledge of our local coasts' for a Hong Kong division of the Royal Naval Volunteer Reserve, and over 60 members expressed their interest in signing up in September 1933.[19]

Similarly, upon his appointment as Commissioner of the Cayman Islands in 1934, Allen Wolsey Cardinall founded both the Cayman Islands Yacht and Sailing Club and an annual sailing Regatta from January 1935. Cardinall wanted to improve the economic condition of the islands, which were sustained predominantly by turtling, sharking, rope-making and boat-building, and he suggested that the Royal Navy recruit there. The Admiralty concurred that 'as a potential source of seamen for the Auxiliary Patrol Service, these ready made sailors seem to justify every encouragement', and sent a warship annually to attend the regatta and provide two sailors to accompany each schooner. Furthermore, the Admiralty loaned a trophy for the race winners, hoping that 'the presentation of such a prize would stimulate interest in the Royal Navy that might be invaluable in the event of hostilities'. Indeed, during World War II, 201 Caymanians volunteered for the Trinidad Royal Naval Volunteer Reserve and another 800 served in the Merchant Navy, representing two-thirds of the islands' adult male population.[20]

Fig. 4.3. Trophies won by HMS *Hawkins* at Weihaiwei in 1923

The Boy Scout movement was also part of the cultural mobilisation in support of British imperialism. The very first edition of Robert Baden-Powell's *Scouting for Boys* advised Scouts to collect postcards of naval vessels to acquire an appreciation of imperial defence issues and an ability to recognise potential enemy warships. Commissioner Cardinall formed two troops of Sea Scouts and Cub packs in Grand Cayman's George Town and West Bay. Like colonial yacht clubs, they prepared aspiring volunteers for naval service, such as the future Petty Officer Harry McCoy, who 'left to join the Navy [ . . . ] that had been my life's ambition as a boy [ . . . ] I was a Sea Scout, and being a Caymanian, I already had the salt water in my blood [ . . . ] so I volunteered'. The Sea Scouts were invited aboard visiting Royal Navy warships during the regatta, introducing them to British naval culture. When World War II broke out, Senior Sea Scouts in Trinidad performed signalling duties until they were absorbed properly into the communications branch of the local RNVR.[21]

In Kenya, annual camps were also held at Kilindini naval base during the 1950s for local Sea Scouts, Nairobi Sea Cadet Corps and Mombasa Sea Rangers. There they would go to sea in Her Majesty's East African Ships (HMEAS) *Rosalind* and *Mvita*, and were given instruction in sailing, boat pulling, seamanship, helmsmanship, signalling and rifle shooting. A sporting legacy was left in the region by the visiting cruiser HMS *Enterprise*

Fig. 4.4. Cricket practice in the Trinidad RNVR

in April 1926. A team of 30 crew members toured Kenya and Uganda, playing several local teams over a fortnight before taking part in a Royal Navy versus Kenya boxing tournament at Nairobi. This was followed by matches in Zanzibar, Dar-es-Salaam, Tanga and Aden. With missionary spirit, the ship's chaplain and tour manager, Mr Harold E. Stevens, departed with the belief that,

> we made rugger history by our up-country tour, being the first Naval team to venture so far into the heart of Darkest Africa [ . . . ] we hope it will be remembered when the various districts come to play for the cup which we presented.

The crew donated a silver goblet to the Rugby Football Union of Kenya, though being an ardently amateur organisation it first sought permission from the Rugby Football Union in Britain before awarding the trophy to the annual winners of the inter-district championship. The Enterprise Cup continues to be contested to this day.[22]

This sporting tradition was inherited by the Senior Service's local heirs. In 1952, the Royal East African Navy (REAN) constructed two football pitches and a .22 shooting range at Peleleza in Mombassa, while every member of the force participated in track and field, football or boat sailing on a weekly basis. This led to one rating winning the Victor Ludorum Cup at the 1953 Coast Athletic Sports competition, while a Sick Berth attendant was selected to represent Coast Province at the 1959 Colony Athletics in Nyeri. It also produced a number of African coxswains, who enabled an annual sailing regatta to be held on Trafalgar Day, providing a testing-ground for prospective petty officers. In June 1957, the REAN won a three-nation regatta featuring the visiting East Indies flagship, HMS *Ceylon*, and INS *Delhi*, flagship of the Indian Navy.[23]

REAN athletics competitions were held annually from 1953 at Peleleza recreation ground, becoming 'popular [ . . . ] with many relatives of naval personnel' and the general public who spectated. Boxing, shooting and football were all 'very popular with the African ratings', with matches held throughout the year against local clubs and visiting ships. In their first season competing in the local league, the REAN football team won seven matches, drew three and lost five, while Leading Seaman Ali Said competed in the national Intermediate Boxing Championships in Nairobi in November 1961 and won 'most promising boxer' at the De Witt Trophy competition. Provincial information officers publicised the fact that 'African personnel of the REAN [Royal East African Navy] [ . . . ] are now enjoying greatly improved welfare facilities under a new campaign recently instituted by the Naval authorities'. The navy's efforts went beyond recruitment and training and helped to justify Britain's colonial development, as improvements in the social and economic condition of local ratings, their families and the community illustrated East Africa's 'progress' under paternalistic British rule.[24]

In Malaya, association football became the most popular sport among the peninsula's indigenous population. Beyond formal matches against European administrators and the exclusive society that accompanied them, Royal Navy personnel shared the game with poorer Malays who taught them how to kick a ball barefoot in return. British officials used the game to improve Chinese integration and social engagement in the colony. Football's working-class roots meant that it avoided the social stigma and segregation that accompanied rugby clubs; it was presented as an inclusive and healthy pastime that physically improved men and was equally popular with naval ratings. The opportunities it provided for gambling were also part of its local appeal.

The resurgence of Chinese secret societies, which had become resistance movements against the Japanese occupation, concerned the British in post-World War II Singapore. The Royal Navy created boys clubs to try and curb the gangs' influence and draw male youths off the streets with sports such as football, cricket and softball. These were a 'big draw at first', before the secret societies put pressure on the local Chinese running the clubs and took over. Navy chaplains were used in this paternalistic mission to recruit boys from school, giving them tours of the base HMS *Terror* and providing puzzles and stories in the chapel. Concerts were also organised, but many local boys did not identify with these Western cultural imports, and preferred the Chinese street performances arranged by the gangs. More successful was the Malaya Cup, named in honour of HMS *Malaya*, which in 1921 visited the protectorate that funded her construction and whose name she bore. Its officers donated two cups for 'the furtherance of football of both codes in Malaya', and the popularity of the inter-state association football competition meant that its successor, the Malaysia Cup, is still contested there.[25]

Malaysian officers trained under the British inherited an appreciation of sport's importance in building their navy and nation. Physically gifted officers, such as Karu Selvaratnam, who joined the Royal Malayan Navy (RMN) in 1960, were prized for their sporting attributes. Though he signed up as an artificer apprentice, Karu spent most of his five years' training playing sport. He represented Malaya and Malaysia internationally at cricket, hockey, badminton and athletics, competing in the 1962 British Empire and Commonwealth Games and winning bronze and silver medals for the individual and team 400 metre hurdles events at that year's Asia Games. When he attended a course at the Britannia Royal Naval College in 1966, his sporting pedigree helped Karu culturally assimilate better than Malaysian cadets who 'didn't have the mental strength [and were] physically weak too'. When Thanabalasingam Karalasingam was appointed as the RMN's first Malaysian Naval Chief in 1967, he sought to resurrect the 'old glory' days of naval sporting success, and appointed Karu to oversee its development. An annual 'sports carnival', which began in 1970, was held at KD *Malaya*, the RMN's headquarters. This event grew to over 1,000 participants across 18 events within five years. Sport thus played an important role in nationalising the force; it was a tool for forging identity, pride and unity, crucial in any young navy and nation, but even more so for one drawn from a plural society with a colonial history of ethnic division.[26]

CHAPTER 5

# Colonial Navalism

The demographic and economic growth of the colonies stimulated calls for a greater voice in managing their domestic affairs. Self-government was granted to Canada, New Zealand and Britain's Australian colonies during the 1840s and 1850s, but, with increased political responsibilities, they were also expected to take on a greater share of the empire's defensive burden. Imperial expansion and technological advances increased the cost of the Royal Navy for British taxpayers, while the colonial empire, which derived defensive and commercial benefits from a strong navy, provided no regular payment towards its maintenance. The second half of the nineteenth century was also one of uncertainty for the British Empire and its collective security.[ix] The Crimean War and Indian Rebellion 'illustrated the danger involved in undue dispersion of the armed forces of the Crown, and the necessity for taking steps whereby the overseas people of the empire would provide, at least in part, for their own security'. Britain's volunteer movement, emerging in response to possible French invasion, 'gave a stimulus and an example to the Colonies in the direction of self-defence'. The nautical nature of the British Empire meant that imperial defence was invariably seaward in orientation, and the first Colonial Naval Defence Act (CNDA) was passed in 1865 to allow the white settler colonies to raise volunteer bodies for their local maritime protection. Their geographical isolation prompted Victoria, Queensland, South Australia, Tasmania and New Zealand to take this opportunity to

---

[ix]  See also Andrew Baines, *A History of the Royal Navy: The Victorian Age*.

form small auxiliary naval units, though as the base of the Royal Navy's
Australia Station since 1859, New South Wales had less need for guard
ships and focused more on recruiting a Naval Brigade.[1]

The 1878 Russian war scare precipitated the creation of the Carnarvon
Commission the following year, 'to enquire into the defence of British
possessions and commerce abroad'. This reported in 1882 that, 'we see
no reason why the Australian colonies should not make a moderate
contribution in money towards the cost of that squadron which is
maintained by the mother-country for the protection of interests common
to the Colonies and herself'. Public anxiety about the inadequacy of the
empire's naval defence escalated on 15 September 1884 with the
publication of sensationalist editor W.T. Stead's front-page *Pall Mall
Gazette* article asking, 'What is the Truth about the Navy?' At the end of
that year, Rear Admiral Sir George Tryon was sent to take command of
the Australian Station in a region which was becoming a hotbed of
imperial rivalry; the Netherlands occupied the western half of New
Guinea and France attempted to establish a colony in New Ireland, while
Germany's annexation of north-east New Guinea prompted Tryon's
predecessor, Commodore James Erskine, to secure south-east New
Guinea on behalf of the nearby Queensland colony. The Penjdeh Incident
in 1885 created another crisis with Russia which potentially threatened
Australian shipping, and allowed Tryon to secure local political support
for the creation of a unified auxiliary cruiser squadron by identifying
'himself with all their interests, making those interests his own and always
speaking of "our" coasts, "our" harbours, "our" towns'. That year, Sydney
also constructed two of its own two-torpedo boats, manned by the local
reserve.[2]

A Colonial Conference was consequently convened in 1887 to coincide
with Queen Victoria's Golden Jubilee, where the Australian colonies and
New Zealand agreed to contribute £122,597 annually towards the
maintenance of five *Pearl*-class cruisers, to be stationed in Australasian
waters and rechristened with local identities: HMS *Katoomba, Mildura,
Ringarooma, Tauranga* and *Wallaroo.* Though the Colonial Secretary Sir
Henry Holland thought the contribution 'trifling in comparison' to
Britain's peacetime naval expenditure of over £13 million, the British
government did not regard this as 'a mere bargain between the Mother

Country and the Colonies, but as [ . . . ] the first step towards a federation for defence'. Furthermore, in 1893, the government of New South Wales donated Garden Island in Port Jackson to serve as the Australasian Squadron's base, spending a further £300,000 in constructing buildings and supplying the naval yard.[3]

In contrast to the Australasian colonies, at the 1887, 1894 and 1897 Colonial Conferences, Canada refused to pay 'tribute' to the Royal Navy on the grounds that Britain would sustain the service regardless. Fresh in Canadian memories was the fact that they had received no assistance in arresting American fishing vessels that had violated Canada's territorial waters during 1886–7. The British government had not wanted to risk escalating the situation into an armed conflict with the United States, and so the dominion was left to form its own Fisheries Protection Service comprising lightly armed civilian vessels and their crews. When an Anglo-American war scare broke out in December 1895, the Admiralty avoided increasing its defensive commitment by endorsing an army proposal that Canada should raise and man a force of vessels to patrol the great lakes.[4]

That year, the British Navy League was formed and it campaigned to raise public interest in the Senior Service and the empire's naval defence through pamphlets, newspapers, public lectures, political lobbying, school textbooks, youth organisations and the remembrance of Trafalgar Day. A Toronto branch of the League shortly followed, and one of its founders, the former Royal Navy officer turned lawyer Henry Wickham, argued that Canada needed to retain an ability to secure its own waters before committing to any imperial defence scheme, a view endorsed by the British naval reformers Lord Brassey and Admiral Sir Charles Beresford. In 1903–4, the dominion launched two fast, modern steel vessels armed with rapid-fire guns; CGS *Canada* would protect the Atlantic seaboard, while CGS *Vigilant* policed the lakes.[5]

## Royal naval tours

The Second Boer War, which broke out in 1899, stimulated public displays of naval theatre that reinforced the bonds between the Royal Navy and the colonies. The unexpected success of the Boers' guerrilla warfare prolonged the campaign beyond British predictions and led to the creation of concentration camps and a 'scorched earth policy' aimed at eradicating the Boers' logistical support, all of which galvanised the anti-war lobby in Britain. To shore up support for the war effort at home and in the empire, where soldiers were drawn from India, Australia, New Zealand and Canada, Colonial Secretary Joseph Chamberlain arranged for a royal visit in 1901 to recognise 'the loyalty and devotion which have prompted the spontaneous aid so liberally offered by all the Colonies in the South African War, and of the splendid gallantry of the Colonial Troops'. This would stir up jingoism, project a unified imperial front and deliver a psychological blow to convince the Boers that fighting against such an overwhelming force was ultimately futile.

This was the most extensive royal tour attempted at the time, the first where the heir to the throne was accompanied by his wife, and the Royal Navy played a crucial role in facilitating it. HMS *Ophir* escorted the Duke and Duchess of York to Gibraltar, Malta, Aden, Ceylon, Singapore, Mauritius, Australia, New Zealand, Canada and, of course, South Africa, displaying 'a thousand miles of loyalty' which would spin the 'web of empire', as an official tour publication professed. It was closely followed by both the imperial and colonial press, becoming the subject of books and films in Britain and the dominions. In Durban, *Ophir* was greeted by crowds of around 50,000 people packed into the city's streets, harbour, verandahs and windows, wherever there was a vantage point to view the procession, and causing one stand to collapse beneath the weight of interest. All communities were drawn to the occasion, with 1,790 Indians and 1,055 Africans among the 10,000 Sunday school children present, though their segregation reinforced the racial politics of colonial society. The mood was captured by the *Natal Mercury*, which claimed that the tour was 'making the Colonists realise more than ever they have done before their position as citizens of the great Empire, centred around the Throne of Great Britain and the Empire'.[6]

*Ophir* visited Simon's Town in the Cape Colony the next morning, where, at Admiralty House, the Duke received a group of Boer prisoners of war who had apparently shown 'great interest' in the tour and presented the royal couple with a set of serviette rings and a broach fashioned from Kruger gold coins. Even the Dutch-language newspaper *The Veldt* reflected that:

1. *Attack and capture of the French company of the Indies position at Chandernagore in 1757*

2. *HMS* Resolution *and* Adventure *in Matavai Bay, Tahiti, during Cook's second Pacific voyage (1772–75)*

3. *Perilous Position of HMS 'Terror', Captain Back, in the Arctic Regions in the Summer of 1837*

4. NEMESIS *destroying the Chinese Junks in Anson's Bay, Jan 7th 1841*

One glance at our City streets during the Royal visit, the tumultuous crowding of all the thoroughfares, the universal decorations and illuminations, reaching from Simon's Town to Sea Point, proved that the people of the Peninsula, at any rate, are loyal to the core, loyal to England, and worthy of the Empire's wide liberties and glorious traditions.

The English-language *Cape Times* claimed that the 'wonder and awe with which our dark-skinned fellow-subjects will look upon the face of the grandson of the "great white Queen"' proved 'the contentment and loyalty of the millions of the coloured races who live under our rule'. Even more explicitly, *The Veldt* emphasised how HMS *Ophir*'s visit helped Britain's 'civilising mission', by reminding the African population,

that they are no longer representatives of barbarian tribes, but are subjects of the King, sharers in England's Empire, profiting by the Pax Britannica which prevails in their broad lands, and sharers in the Empire's liberties and privileges. It is to be hoped too that their visit had inspired them with some ambitions to raise their peoples to a higher level, and taught them that by labour alone can they share in civilization's glorious gains.[7]

This tour followed a longer tradition. Albert Edward, Prince of Wales, had earlier served in the Royal Navy with his brother Prince Alfred, when both inevitably toured their mother's colonial possessions on professional deployments, attracting even greater public attention with a weightier imperial symbolism because of their lineage. Alfred first visited the Cape Colony in 1860 as a midshipman aboard HMS *Euryalus*, returning seven years later as the 23-year-old Commanding Officer of HMS *Galatea*. Two years before the Suez Canal would open and divert a huge proportion of international shipping away from the Cape of Good Hope route, Alfred's 1867 visit was intended to reassure Britain's South African subjects that the imperial motherland had not abandoned them. Local residents responded enthusiastically, erecting an arch on the road from Simon's Town to Cape Town, which

had on its top a large steamship, with 'Galatea' inscribed upon it, and a funnel out of which real smoke was made to issue as the Duke passed under. Six little boys dressed as sailors formed the crew, and stood up singing 'Rule Britannia'.

They also composed original verses to commemorate the occasion and reinforce a shared cultural heritage, which transcended the waters that separated them:

> Hurrah for England's Honour,
> The Noblest from the Strand
> Maintains the Glory over the Sea
> Of our Dear Fatherland.

The naval bond between the colony and Britain was strengthened by a seafaring tradition common to both the Cape's European and Malay settlers, many of whom were fishermen. In conveying 'their loyalty and attachment to their Queen', the Malays were especially thankful 'for being permitted to worship according to their own faith', and their freedom from slavery; they celebrated this liberal beneficence of British rule with prominent displays of costume and dancing. The Malay community put on a demonstration of their nautical prowess during HMS *Euryalus'* 1860 visit, in an attempt to earn the respect of the Royal Navy crew and thus elevate their status within the British Empire:

> One of the most novel sights afloat [ . . . ] was a boat rowed by Malays, with one of the priests steering. The Malay colours were hung out at the stern, and the rowers were dressed in full costume. This boat, a very fast one, danced around the *Euryalus* like a dicuk upon the crest of the waves, and the loyal followers of Mahomet cheered their Prince with a will.[8]

Yet such proficiency failed to overcome the race-based prejudices that imperial authority was built upon, and though the 'civilising mission' encouraged others to live by Britain's cultural values, Malay attempts at assimilation were ridiculed to reinforce their position of inferiority: 'At one place stood an old Malay, playing "God save the Queen" on a cracked clarinet, who, quite absorbed as he was in his music, and apparently unconscious of all around him, looked exceedingly comic'.[9]

The launch of a new warship provided an important stage for naval theatre and the performance of imperial rites. It would be christened with a bottle of colonially produced wine in front of delegates from the empire, a tradition which began with the launch of the cruiser HMS *Good*

*Hope* in February 1901. Originally christened *Africa*, the change in name was a public show of gratitude to Britain's Cape colonists. The symbolism imbued in the vessel meant that it was chosen to convey Chamberlain to South Africa after the peace treaty was signed the following year, and his departure was broadcast in picture houses across the empire as a celebration of imperial victory and togetherness that masked Britain's damaged prestige. With a spirit of reciprocity, the Cape offered an annual contribution of £30,000 towards the Royal Navy's upkeep, and formed a Cape Colonial Division of the Royal Naval Volunteer Reserve on 1 February 1905. The colony's women arranged for a silk ensign and gunnery shield inscribed with the names of 99 contributing towns to be presented to HMS *Good Hope*, and, to further add to the cruiser's distinct colonial identity, a male eland called Peter was chosen as its mascot. Natal also donated £12,000 annually to the Royal Navy, and in 1907 it too had a cruiser commissioned in its honour, prompting the gift of a silver trophy 'presented by the people of Natal as a token of their interest in His Majesty's Navy and of their good will towards the ship bearing the name of the Colony'.[10]

The dedication of new vessels to key parts of the empire was an attempt to strengthen the ties of sentiment by aligning colonial identity with that of the Royal Navy. This was particularly important as the imperial bonds were being loosened by political developments in the settler colonies, where federation stimulated a nationalist surge as the new dominions attempted to unify formerly separate societies with distinct cultures and histories under a new collective identity. A swathe of Royal Navy vessels followed HMS *Good Hope*, as Britain addressed these internal challenges to the empire's integrity at the same time as it responded to the external threat of German naval armament. With the introduction of the *King Edward VII*-class to the fleet between 1903 and 1905, it was 'desirable' that 'the Imperial idea [ . . . ] runs through the naming of this class of battleships', as the First Lord of the Admiralty, Lord Selbourne, explained when launching HMS *Hindustan*:

> The idea of the class was to group round the 'Sovereign' the component parts of the Empire. We began with those two great colonies that had done so much for us fighting side by side in South Africa – the 'Dominion' [of Canada] and the 'Commonwealth' [of Australia]. We then passed on to that wonderful colony,

Fig. 5.1. Admiral Sir Percy Scott with Peter, the eland bull mascot of HMS *Good Hope*,
1908

smaller than the other two, which sent no fewer than ten separate contingents
of men [ . . . ] 'New Zealand'; and then to the gem of the British Empire [ . . . ]
'Hindustan'. The last three ships of the class would be the 'Africa', and then,
coming home, they would have the 'Hibernia' and the 'Britannia'.

The Admiralty ensured that the imperial and colonial press reported the
launch of these symbolic vessels. Trans-continental telegrams were
exchanged and colonial representatives donated gifts so that a physical
piece of the territory would be carried by its naval namesake to provide a
source of pride and affinity for its people.[11]

## The 'yellow peril'

At the same time as Britain was struggling to defeat the Boers in South
Africa, 2,207 Royal Navy personnel were fighting the rebellious Society
of Righteous Harmonious Fists in China. The 'Boxers', as they were more
commonly known, challenged the growing foreign influence in China,
particularly Christian missionary work and the granting of concessions.
Joined by members of the Imperial Chinese Army, they attacked railways,
mission stations and foreign holdings in Tientsin and Peking, causing local
British representatives to request in May 1900 that the Royal Navy step in

Fig. 5.2. A Naval Brigade entraining during the Boxer Rebellion

to quell the uprising. Displaying a remarkable degree of objectivity, even the China Station's Commander in Chief, Vice Admiral Sir Edward Seymour, admitted that British imperial policy since the Opium Wars had contributed to this volatility: 'The general history of our dealings with China has been that we have forced ourselves upon them and into their country. I believe that we are too apt to forget this'.[12]

On 27 June 1900, the First Lord of the Admiralty, George Joachim Goschen, requested through Chamberlain that the Australian colonies send several shallow-draught vessels of their Auxiliary Squadron, which could operate up-river and arrive in China faster than those from Britain. Keen to encourage the colonies to contribute more to imperial matters, Chamberlain was also conscious that the apparent 'omnipresence of available British resources' would provide a boost to British prestige, which had been dented by the Boers. South Australia responded by offering the gunboat *Protector*, and though the number of volunteers diminished once it was revealed that they would serve ashore and not afloat, New South Wales still provided a Naval Brigade of 250 bluejackets, while another 200 came from Victoria. Enthusiastic crowds sent the men off from Sydney and Melbourne, and they arrived in Hong Kong on 27 August. With south China and the Yangtze valley remaining quiet, *Protector* was engaged in survey work and ferrying dispatches in the Gulf of Pechili. The New South Wales Brigade were deployed to Peking, while the Victorian contingent policed Britain's concession in Tientsin and participated in October's Poatingfu expedition. In an early example of coalition warfare, the Royal Navy contributed men to an 'International Naval Brigade' representing the major imperial powers with interests in China (Britain, France, Germany, Japan, Russia, Italy, Austro-Hungary and the US), and helped relieve the Peking Legations from a siege reminiscent of Lucknow. A total of 259 Royal Navy personnel died in China during the rebellion, a higher casualty rate over a shorter period than in South Africa, which caused Vice Admiral Seymour to decry the lack of British recognition given to their efforts and the absence of a medal clasp for Tientsin's defence, 'especially when it is compared to what some clasps were given for in another continent at about the same time'. As a ditty penned by 'Naval R.K.' put it: 'only those that fought Kruger/Are allowed to give themselves airs':

> We got no Tam o'Shanter caps,
> No chocolates in a box,
> No knitted stockings came our way,
> But we gave the Boxers socks.[13]

For the Secretary to the Admiralty Hugh Arnold-Forster, the selfless willingness of Australians to step up and help Britain shoulder the 'white man's burden' in China provided proof of their progress towards nationhood, and set an example that the rest of the empire should follow, including the senior dominion, Canada. Imperial ideologies were also inherited by the white settler colonies, with the 'barbarities' committed by the Boxers becoming racialised as Chinese traits more generally, morally justifying the war as a righteous crusade in defence of Christianity and civilisation. It also fuelled fears of the 'yellow peril' and hordes of Asian immigrants eroding Australia's Anglo-Saxon character from within unless white colonial troops stepped up to defend the British Empire's honour overseas and prevent it from being weakened by an over-reliance

Fig. 5.3. Naval Brigade repairing the line after the Boxer Rebellion

on the Indian Army. While supporters believed that Australia's involvement provided its naval personnel with a valuable opportunity to learn from modern European armed forces, those opposed argued that Australia's continental defence required sailors with seagoing experience and not knowledge of fighting imperial wars on foreign shores, and that militarisation sapped the country of money and manpower which could be used to settle its vast open lands and defend them against Asian invasion.[14]

In 1901, Australian federation saw the old colonial navies merge to form the Commonwealth Naval Forces. The country continued its financial contribution towards the Royal Navy's Australasian squadron, but Lord Selbourne wished to revoke the original 1887 agreement which gave the Australian government the right to block the cruisers' redeployment outside local waters:

> The safety of the Empire can only be secured by seeking out and destroying the ships of the enemy wherever they are [ . . . ] The cruisers which are the subject of the present agreement are small and are rapidly becoming obsolete. The Australians press, and naturally, that the ships to replace them should be first and second class cruisers of the newest type. Under no circumstances will I consent to an arrangement by which the hands of the Admiralty would be tied

in respect of the orders they might wish to give to such ships on the outbreak of war.[15]

A powerful advocate for greater empire unity, Joseph Chamberlain saw cooperation in its naval defence as a step towards his ultimate goal of imperial federation. The 1902 Coronation Review was expanded specifically to impress the empire delegates attending the Colonial Conference in the weeks prior. At the conference, Chamberlain appealed for dominion

> assistance in the administration of the vast empire which is yours as well as ours. The weary Titan staggers under the too vast orb of its fate. We have borne the burden for many years. We think it is time that our children should assist us.

That same year, the prominent geographer and academic Halford Mackinder published *Britain and the British Seas*, where he argued that if the empire were to survive the challenge of powers possessing greater natural resources such as Germany, Russia and the United States, its 'daughter nations' must grow 'to maturity' and contribute towards expanding the Royal Navy 'into the Navy of the Britons'. Though Lord Selbourne was 'doubtful if the time has yet come to ask for any contribution from the Transvaal or Orange River Colonies or from Rhodesia', Australia and New Zealand did increase their annual subsidy to £240,000 in exchange for a more modern and effective squadron, while it was agreed that a vessel should be devoted to training local seamen. The Admiralty reclaimed the right to dictate the cruisers' wartime deployment, prompting Australian criticism that this constituted taxation without representation. The naval issue provided a cause for cultivating an Australian national identity, particularly after First Sea Lord Admiral Fisher's 1904 reforms decommissioned several cruisers of the Australasian squadron without replacing them. The Director of the Commonwealth Naval Forces, Captain William Rooke Cresswell, with the support of the nationalist Labour Party and newspapers such as the Sydney *Bulletin* and Melbourne *Age*, argued that the Royal Navy could not be relied upon to put Australia's interests first and thus the country needed its own separate navy.[16]

Proof of this was seen in Britain's 1902 treaty with Japan, which precipitated the withdrawal of Royal Navy vessels from the Pacific – notably the battleships of the China Station – to meet the German challenge in Europe. This effectively left the naval defence of Australia and New Zealand to the Imperial Japanese Navy. 'Yellow peril' fears were exacerbated by Japan's imperialist ambitions in the region and its comprehensive victory in the Russo-Japanese War of 1904–5. The Russian fleet's annihilation at the battle of Tsushima, the first time a major European power had been completely routed by an Asian navy, shook white assurance in their racial superiority, and prompted Senator Sir George Foster Pearce to warn Australia in November 1905 about the threat from the Rising Sun:

> Japan has shown that she is an aggressive nation. She has shown that she is desirous of pushing out all around. What has always been the effect of victory and conquest upon nations? Do we not know that it stimulates them to further conflict? To obtain fresh territory? Has not that been the history of our own race? Is there any country that offers such a temptation to Japan as Australia does?

To restore confidence, Australian Prime Minister Alfred Deakin invited Theodore Roosevelt's 'Great White Fleet' to visit Australia as part of its 1908 world tour. The sight of the white-painted hulls provided a striking metaphor for Australians who feared the dual 'yellow perils' of Japanese imperialism and Chinese migration, and saw the development of their own navy as a means of preserving the country's racial purity and enforcing its 'White Australia' policy. The demonstration of Anglo-Saxon solidarity provided 'an armed assertion that the White Race will not surrender its supremacy on any of the world's seas'. Australians received their American guests with 'almost hysterical heartiness'; in Sydney, half a million people from a population of 600,000 turned out to welcome the fleet's arrival on 20 August 1908, more than had celebrated federation. Deakin appealed to the masses gripped by naval fervour and declared his desire to see Australia follow the US example, growing from British colonies to nationhood and becoming a naval power in its own right: 'We live in hopes that from our own shores some day a fleet will go out not unworthy to be compared in quality, if not in numbers, with the magnificent fleet now in Australian waters'.[17]

Fig. 5.4. The Great White Fleet leaving Sydney Heads

Australia took its first step towards this in the aftermath of the visit when it ordered two *River*-class destroyers for the Commonwealth Naval Forces. In Britain, the naval writer Archibald Hurd reflected on the diverging strategic outlook that was prompting the dominion to 'go it alone':

> The British peoples whose shores are washed by the Pacific Ocean are becoming increasingly dominated by the fear of the 'yellow man' [ . . . ] This fear, and not the growth of German naval armaments [ . . . ] is definitely and rapidly moulding the destinies of these Dominions on the other side of the world.[18]

Their location on the Pacific also left British Columbians feeling threatened and isolated and keen to follow the path set out by Australia:

> If permitted, Oriental peoples would swarm over North America with surprising rapidity [ . . . ] that which Australia does Canada should do [ . . . ] Canada on the Pacific is close to the Asiatic peril, and while from our physical conformation our defence may always require to be a maximum of militarism, yet without a large minimum of navalism Canada, if attacked by sea, would present a sad and sorry spectacle.[19]

C.F. Hamilton of the Toronto branch of the Navy League also called time on the British idea of subsidising the Royal Navy, and suggested that Canada needed to form a torpedo flotilla to take care of its own naval defence. This had become an issue of 'national self-respect', in which 'we should be doing the work ourselves, and not hiring Englishmen to do our sea-fighting for us'. Evoking the language of Australian navalists, he argued that Canada should 'keep our defence contribution in our hands, and avoid all approach to the danger of taxation without representation'. When Roosevelt suggested that he wanted the 'Great White Fleet' to visit Victoria and Vancouver, it again provided a fillip for these aspirations by drawing attention to Canada's ultimate inability to prevent a foreign fleet from doing what it liked in the country's territorial waters. On 15 May 1908, the Canadian Rear Admiral Charles E. Kingsmill was transferred home with the task of creating a naval militia. He had already acted as the Royal Navy's Ambassador to Canada in commanding HMS *Dominion*'s visit to the country, when, like their Cape cousins, Canadian women donated a silk ensign to be flown aboard 'their' battleship. Quebec's centenary in July 1908 provided an opportunity to raise the public's naval awareness, with the new battlecruiser HMS *Indomitable* joining CGS *Canada* in the St Lawrence river for the celebrations.[20]

## The Dreadnought Crisis

On 16 March the following year, a budgetary debate in Westminster created the 'Dreadnought Crisis', in which the First Lord of the Admiralty, Reginald McKenna, warned that at the existing rate of capital ship construction the Royal Navy would lose its two-to-one advantage over Germany within three years. The danger could be alleviated with help from the empire, which still contributed disproportionately little towards imperial naval defence.

New Zealand stepped up by offering to fund the construction of a Dreadnought for the Royal Navy. In contrast to Australia, New Zealand's Prime Minister, Sir Joseph Ward, favoured 'one great Imperial Navy with all the Overseas Dominions contributing either in ships or money'. New Zealand lacked the resources of the larger dominions to maintain a navy of its own, and Ward feared that 'New Zealand's maritime interests in her

Table 5.1 Imperial defence expenditure, 1908–10

| Year | Country | Population (million) | Military expenditure (£million) | Naval expenditure (£million) | Total expenditure (£million) |
|------|---------|------|------|------|------|
| 1909–10 | Great Britain | 44.539 | 27.459 | **35.143** | 62.602 |
| 1907–8 | Canada | 6.154 | 1.359 | **0.099** | 1.458 |
| 1907–8 | Newfoundland | 0.234 | 0.000 | **0.003** | 0.003 |
| 1907–8 | Australia | 4.222 | 1.025 | **0.272** | 1.297 |
| 1908 | New Zealand | 1.021 | 0.193 | **0.100** | 0.293 |
| 1908 | South Africa | 5.474 | 1.258 | **0.085** | 1.343 |
| 1908–9 | India | 294.317 | 20.071 | **0.482** | 20.553 |
| | Colonies | 38.871 | 1.155 | **0.000** | 1.155 |
| | Total empire | 394.83 | 52.522 | **36.184** | 88.704 |

own waters and her dependant islands in the Pacific would [ ... ] be almost entirely represented by the Australian Fleet'. Whereas Australia saw naval autonomy as a means of ensuring its national survival, New Zealand believed that strengthening its bond with the Royal Navy would achieve the same goal by preventing the country from becoming politically and strategically subordinate to Australia.[21]

Assistance also came from the dependent empire, with Britain's protectorate in Malaya offering to fund a Dreadnought in 1912. Sultan Idris of Perak provided the inspiration and drive behind the idea, as it gave him an opportunity to improve his own political position by earning Britain's endorsement and impressing his own Malay subjects,

> [the ship] will enhance in their eyes the wisdom, and foresight of their rulers whose cordial and unanimous support enabled Sultan Idris' proposals to be carried to fruition and it will strengthen their devotion to His Majesty the King Emperor who personifies for us all both the unity and strength of the British Empire and the splendid traditions of the Imperial Navy.[22]

Like the dominions, Malaya saw the navy in both imperial and nationalistic terms. By proving that the country had matured economically and politically, it was entitled to greater respect within the family of the empire beyond the traditional boundaries of racial hierarchy: 'It is the symbol of a feeling of individual partnership on equal terms amongst all who enjoy the protection of British rule'. This expectation was idealistic for the time, and while the Admiralty and Colonial Office readily accepted the proposal, the unity it created was instead between navalists and disarmament supporters who decried 'the dangerous character of the precedent' offered by a supposedly dependent territory. Following the tradition of HMS *Good Hope* and the *King Edward VII*-class, the new oil-fired battleship of the *Queen Elizabeth*-class was named HMS *Malaya*, and it stimulated great local 'expectations, anticipations, and excitement' ahead of its first visit to the protectorate. In November 1913, Bombay announced that a number of Indian princes also wished to gift the Royal Navy two or even three Dreadnoughts, but these never materialised, and it was suspected that the offer was a political stunt. Whilst lacking the resources for a Dreadnought, chiefs from the landlocked African Protectorate of Basutoland did fund a fast gunboat for submarine hunting.[23]

Australian newspapers were dominated by the 'Dreadnought Crisis' for days, and even the nationalist Melbourne *Age* joined in the general cry for the country to, at the very least, match New Zealand's offer of a Dreadnought for the Royal Navy as a matter of national pride. In Sydney, members of the public took the lead by forming a 'Dreadnought fund' to raise the necessary capital, but the federal government pressed on with its construction of an indigenous destroyer force.[24]

Admiral Fisher consciously cultivated close relations with the press in order to promote the Royal Navy's cause. He arranged for 1909's Spithead Fleet review to coincide with that June's Imperial Press Conference, and invited journalists from across the empire to observe 144 warships in seven lines stretching across 18 miles. The Melbourne *Age*'s Chief Reporter Frank Fox was one of many seduced by this staggering display of naval theatre, which reminded the young dominion of its position within the imperial family: 'How vast to the Australian eye that power [ . . . ] the ramparts of the Empire – wonderful as manifestations of human energy and human genius, more wonderful as manifestations of pride of race'. The Imperial Press Conference was consequently dominated by discussions of naval defence, filtered back in reports to the dominions. In the conference's opening address on 5 June, the former Liberal Prime Minister Lord

Fig. 5.5. The 1909 Spithead Naval Review

Rosebery stressed that 'it is under the Navy that under the good providence of God, the wealth, the prosperity, and peace of these islands and of the Empire mainly depend', and yet it was 'always an inadequate armada'. Knowing the dominions' reluctance to revisit the old issue of direct naval subsidies, British speakers encouraged the development of local forces paid for and operated by the dominions which would fall under Admiralty direction during wartime. The call from Howell Arthur Gwynne, editor of the London Standard, was 'let us have five fleets and one navy'.[25]

This was followed up at August's Imperial Conference on Defence, where the Admiralty presented the delegates with a suggested programme for the construction and maintenance of four imperial naval units, whose command and training would be coordinated to allow them to integrate with the Royal Navy's main fleet when required. These would be operated by Australia, Canada, a joint British–New Zealand China Station unit and an East Indian unit led by Britain, with each comprising an Indomitable-class battlecruiser, four cruisers, destroyers and submarines. Much more effective than the planned destroyer unit, both of Australia's main parties supported the idea. Canada was less enthusiastic, as its unit was intended to integrate with an 'Eastern Fleet of the Empire' though the country also had an Atlantic seaboard to protect. Instead, it passed the Canadian Naval Service Act of 1910 and ordered two cruisers from the Admiralty to patrol each of its coasts, requiring seconded Royal Navy personnel to operate the vessels and provide training.[26] During World War I,[x] Royal Canadian Navy personnel would be dispersed amongst the ships of the main fleet, with four midshipmen losing their lives aboard HMS *Good Hope* at the Battle of Coronel.[27]

Though it would federate with Canada in 1949, Newfoundland also held dominion status from 1907. Once one of the most valuable territories within the British Empire due to its rich fisheries, the island gained the reputation of being a 'nursery of seamen', with 65,000 of its 250,000 population engaged in working the sea at the turn of the century. Despite Colonial Office pressure to provide for their own defence, moves to establish a rifle volunteer corps were voted down by the local cabinet. Newfoundland's Prime Minister Robert Bond believed that a naval force would be less of a drain on the islands' resources, while attracting greater support from its maritime community. Admiral Fisher opened discussions in 1898, while to promote the idea Governor McCallum toured the island's English Shore the following year with Commodore G.A. Giffard and a cruiser of the Royal Navy's North America and West Indies Station, drawing 300 registrants. The capital of St John's was a major relay in the transatlantic telegraph network and thus vital to imperial defence, an argument Bond made at the 1902 Colonial Conference where Newfoundland offered £3,000 annually for maintaining a local Royal Naval Reserve (RNR) and £1,800 for refitting the drill ship HMS *Calypso*, which the Admiralty provided on concession. Volunteers were enticed by the opportunity of joining Royal Navy warships of the North America Station for a six-month training cruise in winter, a season when local fishermen had little else to do. From this, HMS *Charybdis'* Commodore Giffard judged that the Newfoundlanders were 'a useful and efficient body of men who would be a formidable addition to our personnel'. Membership was restricted to 1,000 personnel until war broke out, when they were assigned to man British and Canadian auxiliary cruisers, and by

---

[x]  See also M. Farquharson-Roberts, *A History of the Royal Navy: World War I*.

Fig. 5.6. Members of the Newfoundland RNR in front of HMS *Calypso*, St John's,
*c.*1910

June 1917, 1,713 Newfoundlanders had joined the Royal Navy, 124 of
whom were killed in action or drowned.[28]

New Zealand's battlecruiser was commissioned in November 1912,
paid for by the dominion and thus it was deemed appropriate that the ship
should also bear the country's name to tighten that bond further. This
meant renaming the existing *King Edward VII*-class HMS *New Zealand*, but
the High Commissioner and Governor-General's suggestion of HMS
*Maori* was rejected by the Admiralty, who considered it inappropriate to
name a capital ship after a 'native' people. The name was deemed
acceptable for two Tribal-class destroyers though, reflecting the imperial
hierarchy which dictated that colonised 'races' should follow Britain and
not lead the fleet. In the end the battleship was creatively renamed HMS
*Zealandia*.

HMS *New Zealand* would not be deployed on the China Station as per
the original agreement, nor would the East Indies receive a similar
battlecruiser from the Royal Navy. British fears following the German
Naval Law of 1912 led Churchill to inform the House of Commons on 26

March 1913 that the vessel would be stationed at Gibraltar, where it could reach the 'Channel in a very much shorter time' than the 32 days it would take from the Pacific. This decision was taken without colonial consultation, and, to show what they had paid for, it was arranged for HMS *New Zealand* to visit the dominion between 12 April and 25 June 1913 as part of an empire tour. By calling at as many New Zealand ports as possible, Captain Lionel Halsey hoped 'it would be the means of creating a wave of Imperial Loyalty'. With five New Zealand officers and Prince Henry of Battenberg among the ship's company, 578,937 visitors turned out to witness their battlecruiser – almost half the islands' population. The local branches of the Navy League helped whip up this enthusiasm and worked with education authorities to ensure that New Zealand's youth received this naval propaganda, distributing free picture cards and transporting 5,000 schoolchildren in Wellington to a naval history lecture by Captain Halsey. The huge parades, parties and dances held in their honour made Halsey feel 'much more like a king than a poor naval officer', and he was showered with kisses by the local women including a 'beautiful Maori princess'.[29]

The previous HMS *New Zealand* had received a Maori war canoe, and a gunnery shield depicting Auckland, Wellington, Christchurch and Dunedin surrounded by Maori weapons, New Zealand birds and flowers, and the colonial arms. This bond was reinforced with new gifts in 1913, including – once again – a silk ensign and jack made by the dominion's women, while in Wellington, a Maori delegation from the Ngatiapa tribe greeted the 'war canoe' and presented a rare Kiwi mat to Halsey, with the following speech:

Brave ones, sons of the great God of War Tutengauaghan [ . . . ] Welcome with power, welcome with thy exultation, welcome with thy awe-inspiring presence, welcome thou canoe of war, the sentry that is for ever wakeful upon the mighty deep and upholder of the British Mana [authority]. Let us view the offspring or fruit of the seed cast upon the waters by our Mantua (father), Sir Joseph Ward. The act of such a statesman proved to be the moving spirit of closer relations between the Dominions and the Motherland, who holds the power of protection and peace over her Dominions which make her stand to-day pre-eminent among nations. Welcome, ye defenders of the fort. It is

Fig. 5.7. Captain Halsey greeting Vancouver civilians during HMS *New Zealand*'s 1913
Empire Tour

expressed by a proverb in Maori: 'The house that stands without the fort is
food for fire.'[30]

This imperial unity was further strengthened by the gift of a Ponamu to
HMS *New Zealand* – an indigenous jade stone which symbolises the Mauri
or life essence, with the Maori believing that '*He manawa ka whitikitia he
mauri ka mau to hono*'. (The heart provides the breath of life, but the mauri
has the power to bind or join.)

    The ship's mascot, a bulldog called 'Pelorus Jack', was donated by a
New Zealander living in Britain, and was named after a friendly dolphin
famous for escorting vessels through the treacherous Cook Strait. This
bulldog became a focal point for expressions of unity throughout the
empire, receiving two silver collars from South Africa after its
government requested that HMS *New Zealand* visit during its 1913
tour, as this 'could not fail to have excellent effect in impressing popular
imagination and quickening interest in the Navy'. Pelorus Jack's bad
temper made him less popular with his shipmates, who welcomed his
untimely death after he fell down the ship's forward funnel in April 1916.

Fig. 5.8. Pelorus Jack, mascot of HMS *New Zealand*

As a member of the ship's crew, he was given a Certificate of Service, Medical History Sheet, Certificate for Wounds and Hurts and even a will, which patriotically requested that:

> My successor should be a bull pup of honest parentage, clean habits, and moral tendencies. But it is my earnest wish and hope that no Dachshund or other dog of [German] extraction be appointed to the said office or be otherwise permitted on board HMS *New Zealand* except only by way of rations or mess allowances for my successor.[31]

As per his last wishes, a replacement bulldog was found, but Pelorus Jack II developed an unfortunate fear of gunfire following the Battle of Jutland, and died whilst quarantined in New Zealand in 1919.

Following HMS *New Zealand*'s departure from the Pacific back to Britain, the dominion's Defence Minister Colonel Allen declared in December 1913 that 'the Chief partner of the 1909 Agreement – namely the Mother Country – has failed to carry out her obligation'. Yet the 1910 Committee of Imperial Defence had admitted that 'to avoid exposing our fleets to the risk of suffering defeat in detail, naval action in remote waters might [ . . . ] have to be postponed until [the] clearing of the situation in home waters'.[32]

Around 400 New Zealanders served with the Royal Navy during World War I; they included William Prize, who transported soldiers of the Australian and New Zealand Army Corps (ANZACs) to Egypt for the Gallipolli campaign, before receiving a Victoria Cross for his actions against the German submarine *U-93* while commanding HMS *Prize*, a Q-ship which masqueraded as a merchant vessel. Though the U-boat was seen going down with its bow sticking vertically out of the water, she limped back to port and provided a description enabling *U-48* to torpedo the *Prize* with the loss of all aboard her. On 23 April 1918, New Zealanders including Lieutenant M.S. Kirkwood, who would receive a Distinguished Service Cross, helped the Royal Navy take the Belgian ports of Zeebrugge and Ostend, which had been used as bases for German U-boats. HMS *New Zealand* also played a notable role in all three of the North Sea's major battles.[xi] She was present at Heligoland Bight on 28 August 1914, before her salvoes disabled and helped sink the German armoured cruiser *Blucher* at Dogger Bank on 24 January 1915. Then, at Jutland on 31 May 1916, as part of Admiral David Beatty's battlecruiser squadron, *New Zealand* landed hits on the German cruiser *Koln* before it was finished off by the flagship HMS *Lion*. A major hit on *New Zealand*'s X turret caused no casualties, which crew members claimed was due to the fact that their Captain was wearing the sacred *tiki* and *piu piu* gifted to the ship by a Rotoruan Maori chief to ward off evil in battle.

On 10 July 1911, the Commonwealth Naval Forces were granted the title of 'Royal Australian Navy' (RAN) by King George V ahead of the launch of its flagship HMAS *Australia*. Donated by the dominion was a silver model of Cook's *Endeavour* on a dish of pearls from Thursday Island, and a silver shield bearing the motto 'Our heritage the Sea' above a map of Australia with the names of its naval founders Dampier, Cook, Bass and Flinders in the corners. On 31 June 1913, the King and Prince of Wales visited the battlecruiser before it departed Britain. In front of cinematographers, members of the press and an official reporter appointed to cover the voyage for the empire's readers, 'Australia's

---

[xi]   See also M. Farquharson-Roberts, *A History of the Royal Navy: World War I.*

Fig. 5.9. HMAS *Australia* arrives in Sydney harbour

Admiral' George Patey was knighted on the quarterdeck, an act not performed since Sir Francis Drake. The dominion already possessed a strong nucleus of naval personnel, thanks to the 1902 training agreement, and the ceremony was followed by an impromptu 'three cheers for Wallaby land' from the crew before 600 Australian expatriates toured the ship. Responsibility for the dominion's maritime protection passed to HMAS *Australia* when she arrived 'home' on 4 October 1913. Huge crowds welcomed her to Sydney Harbour with alternating renditions of 'Rule Britannia' and 'Advance Australia'. This was followed by celebratory dances, dinners, parades, fireworks, carnivals, naval film screenings and commemorative publications which reinforced the expression of imperial unity, including one postcard where the battlecruiser is seen draped in the Southern Cross and Union Jack with the slogan: 'One King, One Fleet, One Nation'. The RAN reverted to Admiralty control when war broke out, seeing service as far afield as the Indian Ocean, East Africa, the Suez Canal, Dardanelles, Mediterranean, North Atlantic and North Sea, while HMAS *Sydney* defeated the German cruiser *Emden* off the Cocos Islands on 9 November 1914.[33]

South Africa only became a dominion with the Union of Natal, the Cape Colony, Orange River Colony and Transvaal in 1910, a year later than the Imperial Defence Conference Agreement. By August 1912, the country's first Prime Minister, General Louis Botha, wished to 'follow in the footsteps of other dominions' by developing a navy in lieu of annual contributions:

> I want our people to realize the importance of our sea-bourne traffic. The value of our sea-borne trade is something like £100,000,000, and is increasing day by day. It requires the attention of every man in South Africa. I want the people of this country to realize that if they are to be responsible for the protection of this country, they must expect to be made equally responsible for the protection of their sea-borne trade [ . . . ] we might wish to have a small navy in our own waters.

As part of the new Union Defence Force, the Natal Naval Volunteers and Cape RNVR were consolidated into a South Africa RNVR on 1 July 1913, financially maintained by the dominion but trained and administered by the Royal Navy. In August, Australia lent its support by having HMAS *Australia* and the cruiser HMAS *Sydney* call at Cape Town and Durban during their journey from Britain. At a banquet held in Parliament House for the ships' officers, the Cape's Administrator Sir Frederic de Waal used the visit to highlight South Africa's naval plight:

> If tomorrow there were no British Navy the liberty of Canada and Australia would be in danger, and the liberty of South Africa would be even more so, because of the important strategic position they occupied on the great trade route to the East. South Africa would be unable to hold its own if an enemy had possession of its ports. It was therefore right that the great question of the British Navy should affect us living in South Africa.

Even the *Volkstem* (*People's Voice*) newspaper representing Afrikaner opinion advocated 'the principal of the South African Navy'. The dominion's Minister of Defence General Jan Christiaan Smuts, also preferred a 'Kleine Vlooje' (little fleet) over monetary donations, but internal political trouble and the outbreak of World War I curtailed the idea. The RNVR(SA) was mobilised and manned local harbour defences, Royal Navy shore establishments, British warships in the North and Mediterranean Seas and fought successful campaigns in German South-West and East Africa. HMS *Good Hope* and HMS *Natal*, which had been named for the former colonies, were both lost in the conflict, inspiring memorials in Halifax and Durban and providing another example of how the Royal Navy left its physical and psychological mark upon the empire.[34]

# CHAPTER 6

# The Empire Rallies

Victory in World War I saw the British Empire reach its greatest extent, with the addition of 13 million subjects and 1.8 million square miles of German and Ottoman territory. These were formally assigned as Mandates by the League of Nations, to be looked after in trust for an indefinite period of time until they had developed enough to govern themselves. At the 1918 Imperial Conference, the dominion Premiers requested that a senior naval officer undertake a tour to advise on the empire's future naval defence. The following year, Admiral of the Fleet Viscount Jellicoe visited India, Australia, New Zealand and Canada, where he endorsed the development of the dominion navies once his grandiose plan for a powerful Far Eastern Fleet featuring eight battleships and eight battlecruisers funded by empire contributions had been rejected. Jellicoe's mission had particular significance for New Zealand. He flew his flag in HMS *New Zealand*, and desired the Governor-Generalship of the Dominion, which he acquired within a year of his survey. In 1921 a Royal Navy New Zealand Division was formally established with the loan of the light cruiser HMS *Chatham* and the training ship and supply depot HMS *Philomel*, a name which was retained by Auckland's naval base at Devonport.

At the 1921 Imperial Conference in London, South African Prime Minister Jan Smuts agreed that his country would cease its annual £85,000 contribution towards the Royal Navy's maintenance, and instead establish its own permanent naval force for undertaking general duties, minesweeping and hydrography, while assisting in the expansion of the local RNVR and Simon's Town naval base. The Admiralty supplied the minesweeping trawlers HMS *Foyle* and *Eden*, which were renamed HMSAS *Sonneblom* and *Immortelle*, while the *Hunt*-class minesweeper HMS *Crozier* was converted into a survey ship and re-commissioned as HMSAS *Protea* on 1 April 1922. The South African Naval Service (SANS) operated as a branch of the Union Defence Force, though it appointed a retired Royal Navy officer, N.H. Ranking, as its inaugural commander.[1]

## The Singapore naval base

The war had left Great Britain, the United States and Japan as the three major naval powers in the world, all with vested interests in the Pacific. Wartime alliances quickly deteriorated into tension and suspicion as the US and Japan embarked on naval construction programmes. In Britain, there was no longer the political or public support, the economic leverage, or industrial capacity to compete with America in warship production. The solution was a policy of naval arms limitation, agreed at the Washington Conference of 1921–2: a ten-year restriction on capital ships by a ratio of 5:5:3, with Britain and the US allowed tonnage of 525,000 each, and Japan 315,000 tons. To ensure American agreement, Britain had to terminate its alliance with Japan, while several vessels had to be paid off, including HMS *New Zealand*, whose 4-inch guns were sent to the dominion to guard Fort Dorset and Godley Head.[2]

Though the 1922 Washington Naval Treaty maintained good relations and naval parity with the US, it decreased Britain's security in the Asia-Pacific region by turning Japan into a potential enemy. It meant increased responsibilities for the Royal Navy, as Britain's possessions in the region had been guarded by Japan. A 1920 memorandum had established the Admiralty's principle of sending the 'main fleet' east in a war with Japan, but no graving docks existed beyond Malta capable of taking battleships possessing anti-submarine bulges, and so the decision was taken, following Washington, to construct a naval base at Singapore, considered 'the naval key to the Far East' by Jellicoe. The necessary land was donated by the Straits Settlements on 3 May 1923 'as a free gift' equivalent to £146,000, while other colonies also stepped forward with contributions; Hong Kong offered £250,000 pounds in 1924, £2 million came from the

Federated Malay States in 1926, New Zealand provided an additional £1 million the following year, while the Sultan of Johore personally financed £500,000 in 1935. Most of these territories participated on the understanding that they were reliant upon the Royal Navy to defend them from major attack; this also included New Zealand, which, unlike its fellow dominions Australia and Canada, did not administer its own navy but hosted a local Division of the Senior Service, maintaining and manning two loaned light cruisers while having operational control of smaller vessels on the station. Yet the contribution sparked controversy, as Harry Holland, leader of New Zealand's Labour Party and opposition, questioned the timing of the announcement on ANZAC weekend:

> The men who fell at Gallipoli died in the belief that they were yielding their lives to win a war that was to end all wars, and, while the church bells were calling people together to honour the memory of the dead, the propaganda was being made for a project that would arouse the fiercest antagonisms between the East and the West and furnish the guarantee of future warfare.

In its defence, *The Press* of Christchurch argued that the base represented a 'bulwark of the Dominions of the South', not Britain, and thus New Zealand had a 'duty' to contribute. Following their election victory, Britain's second Labour government set up a review of the Singapore project and slowed construction while advocating naval disarmament. The naval base's original estimated cost of £13 million quadrupled as its completion was delayed until 1938.[3]

The strategic imperative given to Singapore was the reason why Britain omitted it from Article Nineteen of the Washington Naval Treaty, which established that the 'status quo [ . . . ] with regard to fortifications and naval bases, shall be maintained' east of the meridian of 110° east longitude. This encompassed Hong Kong and Fiji, meaning naval forces had to play an even more fundamental role in the local defence of those colonies, and the long-established policy of not tying warships to bases had to be revised. On 14 December 1922, the Committee for Imperial Defence suggested establishing a RNVR in Hong Kong to compensate for this physical defence restriction. The idea of colonies contributing to imperial defence by raising local RNVRs was suggested to the Imperial Cabinet the previous year. The Washington restrictions forced a new

'one-power standard', where the Royal Navy's strength only equalled that
of the United States, and meant that 'local forces' would have to plug
strategic gaps until its major warships could respond:

> The requirements of a one-Power standard are satisfied if [ . . . ] arrangements
> are made from time to time in different parts of the world [ . . . ] to enable the
> local forces to maintain the situation against vital and irreparable damage
> pending the arrival of the Main Fleet.[4]

Despite this disadvantage, the Royal Navy believed that 'British material
and superior training would produce decisive results against Japan'. Such
arrogance was born from imperial ideas about British racial superiority,
particularly as Japan modernised its navy by following the Royal Navy's
paternalistic example in training and warship construction after the Meiji
Restoration. Racial prejudice also influenced the opinion that naval
reserves were only viable in colonies possessing sizeable European
communities:

> Difficulty may exist in raising local forces of the RNVR in India and in those
> Colonies where coloured races predominate . . . this should not prevent the
> formation of local detachments composed entirely of Europeans, for it will
> probably be found at large and prosperous ports, where a considerable
> proportion of the white community are connected with shipping, and where
> yachting may be indulged in, that service in a naval corps would be attractive.[5]

## The Special Service Squadron

In 1923–4, a 'Special Service Squadron' steamed 38,000 nautical miles
over ten months touring the empire, in the belief that it 'would
undoubtedly arouse much interest and enthusiasm and would facilitate the
successful adoption of any schemes for the successful co-operation in the
Naval defence of the Empire'. The Admiralty maximised publicity by
soliciting British and colonial newspapers, and embarking a cinemato-
grapher to produce a film of the voyage, *Britain's Birthright*, which would be
released on Empire Day in 1925. Also aboard was the journalist V. Scott
O'Connor, who authored the illustrated book, *The Empire Cruise*, also
published in 1925 with a preface by the First Sea Lord and former Colonial

Fig. 6.1. *Hood*'s mascot, Sussie, entertains with a boxing bout

Secretary Leo Amery who stressed that 'the naval problem, like all our other problems, can only be solved by the co-operation of all the partner nations in the British Empire'. To visibly demonstrate this, it was arranged for warships from Australia and the China Station to join the Squadron for its journey from Singapore to the Antipodes, to encourage 'future close association between China Squadron and Australian Navy', while HMAS *Adelaide* stayed on for the Canadian leg of the tour, 'in order to show Canada what the Australians are doing'. *Hood* was accompanied by another Australian ambassador in the form of Sussie, a kangaroo mascot from the people of Sydney, who entertained crew members with staged boxing bouts.[6]

The Special Service Squadron was led by the modern battlecruiser HMS *Hood*, considered 'the largest and most powerful warship in the world [ ... ] a triumph of British engineering skill', the recently-modernised battlecruiser *Repulse*, and five new *Delhi*-class cruisers. They attracted almost two million visitors, regularly drawing crowds that lined 'every foot of deck space, climbed every ladder, adorned every excrescence upon which a man could stand and suspended itself in monkey-fashion where foothold was denied'. In Melbourne, 'every road and pathway was thick, and many families were making a day of it, taking out all the children and hampers of food and bottles of beer. The Bay was dotted with sailing boats'. Carefully stage-managed, with crews under strict instructions to uphold the navy's professional reputation (though seamen such as Albert Edward Scott nicknamed it the 'booze cruise'), this projection of naval theatre was intended to show the dominions 'the best we have got' and inspire awe and imperial loyalty among Britain's colonial subjects, as *Hood*'s Lieutenant C.R. Benstead observed:

> From the native soldier who, when the Squadron came up harbour, seized his officer's arms and pointed at the Hood crying: 'De man dat made dat de ship him god,' to the white-robed dignified figure [ . . . ] who boarded us and, in amazement, rolling his eyes so violently that, losing his balance, he sat down – all were impressed.[7]

*Hood*'s searchlights stretched far inland, which 'unnerved the more-than-ordinarily superstitious among the tribes dwelling beyond the mountains' of Sierra Leone, while for Fiji's chiefs the effect was 'amazing'. In South Africa, the Governor-General suggested that the squadron's visit to the Cape might produce the politically 'valuable effect' of reinforcing British hegemony, given that 'large numbers of Dutch farmers and other inland residents visit[ed] Mossel Bay'. While intended to foster unity, the tour also highlighted the empire's cultural differences,

> we were invaded by Freetown negroes who gaped in open-mouthed astonishment at a ventilating fan; by the bearded Boer farmers from the South African veldt; by white-robed Mohammedans from Zanzibar who, at sunset, spread their prayer-mats upon the quarterdeck and reverently knelt to invoke the protection of Allah.

Britain's technological and industrial superiority reinforced its imperial leadership, as O'Connor highlighted by contrasting 'our masses of steel and iron, our hungry guns and reservoirs of the most deadly explosives,

Fig. 6.2. Men from the Special Service Squadron with 'native hunters'

the discipline and manhood of an English battleship', with the spectacle of Ceylon's inhabitants, 'frail slight beings born of indolent seas', depicted washing the Royal Navy's laundry in *Britain's Birthright*. Ethnographic shots were used in the film to emphasise the relative primitiveness of colonial subjects, and the importance of British rule to their progress and development. The narrative draws attention to the misleading notion that 'when Singapore was purchased for Great Britain in 1819, it was an utterly deserted island [ . . . ] To-day it is a great modern city, boasting fine public buildings', while 'Kingston shows striking contrasts between the bustle of a modern commercial city [ . . . ] and the primitive chaffering of its negro markets'.

Following the film's release, *Kinematograph Weekly* claimed that *Britain's Birthright* 'shows in a manner never before attempted what a mighty confederation the British Empire is'. The magazine suggested it was a 'picture that every schoolboy and girl should see, since the flag needs "showing" nearly as much in England as it does in the Colonies', and it offered them 'a wide vision and a better idea of the extent and possibilities of the Empire than any number of books'. Royal Navy vessels continued to conduct empire tours, with *Repulse*'s sister ship, *Renown*, visiting Gibraltar, Malta, Mauritius, India, New Zealand, Fiji, Canada and Jamaica, as it conveyed the Duke of York for the opening of Australia's

Fig. 6.3. 'Samoan native boats' row out to the Special Service Squadron

Parliament House in 1927. Yet the navy did not always receive a patriotic welcome, particularly as colonial nationalism began to express itself more vociferously, and *Renown*'s 1921–2 visit to India with the Prince of Wales provoked strikes and rioting in the city of Bombay.[8]

6.4 HMS *Renown*'s non-commissioned officers visiting the Taj Mahal

## Interwar challenges

The end of World War I meant demobilisation for many Royal Navy personnel, and several soldier settlement schemes were instituted to induce ex-servicemen into bolstering the white populations of Britain's overseas possessions. Kenya became '*par excellence* the retired officer's colony' after introducing such a scheme in 1919, assigning over two million acres of land to increase its white-populated area by a third. Not a single European resident over the age of 26 had been born in Kenya, and it was hoped that this 'human sea' of settlers would strengthen the colony economically and politically by bolstering the ranks of its white minority in the face of African unrest, which had grown in reaction to World War I. While the majority of immigrants came from the army, 30 former naval officers settled in Kenya (plus many more in other parts of the empire), including D.B. Crampton, who rather poetically named his retirement farm 'Admiral's End'. Thus even after leaving the service, naval personnel saw themselves aiding Britain's imperial mission, though some came into conflict with the motherland over the best way to advance it. The 1922 Wood-Winterton Agreement between the Colonial and India Offices opposed Indian segregation and any restriction of immigration in Kenya; it also ensured that 10 per cent of the electoral roll were Indian, and reserved four legislative seats for them. Kenya's colonists, while 'small in numbers, were strong in military experience', and consequently orchestrated 'a rebellion which aimed, not at breaking away from the Empire, but at remaining in it'. They saw themselves as defending the true interests of the British Empire having been betrayed by distant ministers, and they adopted the motto 'For King and Kenya', while rebel council meetings ended with bombastic renditions of 'God Save the King'. The Royal Navy despatched the cruisers HMS *Southampton, Cairo* and *Colombo* to the region to dissuade their own people from escalating the situation, until the 1923 Devonshire White Paper settled the European–Indian conflict by declaring the paramountcy of African interests, which would, in the end, undermine white British rule in Kenya.[9]

Britain was not the only country allocated Mandate territories to administer by the League of Nations, as the dominions retained control of the German colonies their forces had captured during World War I. New

Zealand's heavy-handed attempts to 'modernise' Samoa alienated resident Europeans and mobilised the Mau movement, or *O le Mau a Samoa* (the firm opinion of Samoa), which pursued a passive resistance campaign calling for self-government and a 'Samoa for Samoans'. To bolster its own colonial authority, between 1928 and 1930 the New Zealand Government deployed HMS *Diomede* and *Dunedin* to the islands with a Royal Marine contingent, despite them technically being Royal Navy vessels containing a large number of British personnel. While naval parties arrested around 400 Mau in their first action, a lack of holding facilities saw prisoners detained in the Malinu Peninsula by little more than a wire, meaning they could come and go as they pleased: 'if you went down to their quarters during the night you wouldn't find anyone there, they got through under the nets, through the mud and had gone home. In the morning they were always there for breakfast'. Naval personnel guarding Mau prisoners were given a rifle and bayonet but no bullets, with authorities hoping that the impression of firepower would be enough to deter challengers and maintain their control; a metaphor for British imperialism elsewhere. Commodore Geoffrey Blake, the officer commanding the naval forces from HMS *Diomede* and *Dunedin*, advised the local police to 'use their batons more freely' in preserving order, as a situation could quickly spiral out of control if men carried firearms. Unfortunately, his warning was not heeded, and on 28 December 1930 violence broke out during a rally calling for the return of exiled Mau. When members of the crowd resisted police arrests, warning shots were fired, causing a riot in which a policeman was beaten to death. Retreating to the station, a sergeant fired further warning shots with a machine gun, while three police men panicked by firing their rifles at the crowd, injuring 11 Mau and killing eight of them, including High Chief Tupua Tamasese Lealofi III, who had been trying to calm the situation.[10]

The economic problems of the 1920s and 1930s took their toll on the Royal Navy, which had already had its two-to-one budgetary superiority over the army reversed, and its percentage of national expenditure dropped dramatically. Yet the battleship remained central to the Admiralty's strategic doctrine into the 1930s, meaning that auxiliary fleet operations such as anti-submarine warfare, minesweeping and local patrolling were neglected, despite the security of colonial harbours being vital to a roving fleet strategy. In 1931, *Brassey's Naval and Shipping Annual* painted a bleak assessment of the Royal Navy, with 'less money is spent on it; there are fewer ships in hand; [while] replacements are not being provided for as they become due'. This sentiment was echoed to the Committee on Disarmament in April that year, with Britain having disarmed most whilst investing least in modernising its ships, coastal defences and auxiliary services; it was considered that 'both absolutely and relatively the strength of the Royal Navy had declined'. By 1933, British naval spending was well under the record low set in 1924, at just 6 per cent of government expenditure.[11]

The Admiralty increasingly looked to devolve naval duties to colonial reserve units. Lower colonial wages, subsidised by their legislatures, presented potential relief to the Royal Navy's strategic, financial and manpower deficits. Director of Plans, Captain W.G. Egerton, argued at the Overseas Defence Committee meeting of 23 May 1927, that the best way the 'British Colonies could co-operate in the naval defence of the Empire [ . . . ] was by the organisation of local minesweeping forces, manned by local branches of the Royal Naval Reserve or Royal Naval Volunteer Reserve':

> The most important aspect of this question was the organisation of local minesweeping services to ensure that sporadic minelaying would not bring the overseas trade of the Colonies to a standstill, and thereby dislocate the steady flow of shipping in which the empire depended for its very existence.

Having learned from past disagreements with the dominions, the Admiralty had 'no intention of plunging into the old controversy of direct money contributions <u>versus</u> local defence forces', and a new Colonial Naval Defence Act was passed in 1931. In relinquishing these responsibilities, the Admiralty admitted that its resources were too stretched to secure colonial trade, as its own minesweepers 'would be required elsewhere [ . . . ] for keeping the approaches to the Fleet anchorages clear and for safeguarding the approaches to the home ports'. Britain's home defence was still placed above that of the colonies, despite rising tensions in the East.[12]

Japanese expansionism asserted itself with the invasion of Manchuria in September 1931. More worrying was the short war in January to March 1932 between Japan and China's Nationalist Army in Shanghai, which threatened Britain's economic interests there and stirred fears of imperial retreat:

> Preliminary measures could be adopted – such as rupture of diplomatic and economic relations – but in the end Japan can only be checked by force. Ultimately we will be faced with the alternatives of going to war with Japan or retiring from the Far East. A retirement from the Far East might be the prelude to a retirement from India.

'Britain's prestige had to be maintained, for it was on this national asset that possibilities for profitable trade and financial gain depended'. But here, Britain 'lived on illusion. She was conscious of being over-extended and felt vulnerable in the face of actions from restless powers'. By June 1939, the country's prestige had diminished to the point where Japan blockaded Britain's Tientsin concession and publicly demanded it renounce the Chinese Nationalists within. A Royal Navy fleet could not be spared from Europe because of German bellicosity, leading Lord Chatfield to comment that Japan, 'trading on our relatively weak naval position, was insulting British nationals in Tientsin, in a manner that would have made a Georgian or Victorian statesman issue violent ultimatums'. The Manchurian, Shanghai and Tientsin crises demonstrated how unprepared Britain was for an Asian war, and a review of military expenditure in September 1933 gave short-term priority to improving the defence of its eastern possessions, until threats emerged in Europe.[13]

France and Italy refused to limit their fleets in 1930, threatening Britain's position in the Mediterranean, while the 1933 Defence Requirements Committee identified Germany as a potential enemy. The Great Depression caused South African defence expenditure to be cut heavily, hitting the country's Naval Service hardest and seeing its administration revert to the Commander in Chief of the Royal Navy's Africa Station on 4 November 1932. On 30 April the following year, HMSAS *Protea* was paid off as the SANS' complement was reduced. It ceased being a seagoing force on 31 March 1934, when the remaining minesweepers were returned to the Royal Navy, leaving a skeleton shore-based unit of just four officers and 12 ratings. At the London Naval Conference of 1935–6, First Lord of the Admiralty Lord Monsell admitted that the Royal Navy could not uphold its 'world-wide responsibilities [ . . . ] for that would mean denuding its home waters of naval defence'. The reality of Britain's overstretch came to the fore during the Abyssinian Crisis of 1935, when Italy invaded the last independent African kingdom. The League of Nations' credibility was undermined by Britain's reluctance to use the Royal Navy to enforce sanctions against Italy, fearing Japan might take advantage of a Mediterranean distraction to expand into Britain's Asian colonies.[14]

Following a 1936 suggestion from Churchill, the Admiralty decided in May 1937 to recruit Royal Navy personnel in the dominions, but was blocked by Australia and New Zealand who feared the effects on their own forces. The Chiefs of Staff Committee that year admitted that they were unable to 'foresee the time when our defence forces will be strong enough to safeguard our territory, trade and vital interests against Germany, Italy and Japan simultaneously'. Admiral of the Fleet Lord Chatfield reaffirmed this bleak forecast for the empire's naval defence the following year:

> Imperially we are exceedingly weak. If at the present time, and for many years to come, we had to send a Fleet to the Far East, even in conjunction with the United States, we should be left so weak in Europe that we should be liable to blackmail or worse.[15]

### Colonial armament

These geopolitical pressures forced Britain to abandon its reservations towards training and arming organised bodies of non-European sailors. Between 1933 and 1936, indigenous naval units were established in Hong Kong, Singapore, the Gold Coast and Ceylon, while the auxiliary Royal Indian Marine became the Royal Indian Navy (RIN) in 1934. In June 1938 the Overseas Defence Committee conceded that British forces were 'barely sufficient to meet the numerous threats which confront the Empire in all quarters of the world', and beseeched that 'every single Dependency, from the smallest and least well-endowed to the largest and most wealthy, has its own part to play in Imperial Defence'. It was argued that the colonies could contribute to empire security as a whole by taking responsibility for their own local naval defence, allowing the Royal Navy's limited resources to be redeployed to meet greater threats elsewhere:

> At every defended port local naval forces have to be maintained for minesweeping and other local defence duties. For these purposes Naval Volunteer Reserve units are particularly valuable, since they enable a corresponding economy to be made in regular naval personnel, who can be more profitably employed in other duties requiring a higher standard of technical training.

By the end of 1938, Gambia and Zanzibar had also established their own Naval Volunteer Forces, while Penang hosted a Malayan branch of the RNVR. They were joined by Trinidad, Nigeria, Sierra Leone, Kenya, Tanganyika and Fiji in 1939, Burma in 1940 and Mauritius in 1941. A full-time Royal Navy (Malay Section) was also formed at the start of World War II, providing Malay seamen to relieve British ratings aboard Royal Navy vessels deployed to the region. British imperial leadership was preserved in these colonial reserves by a colour bar, which prohibited Asian and African ratings from becoming commissioned officers.[16]

Global rearmament in the late 1930s saw several new warship orders placed for the Royal Navy, which were promoted in imperial terms. The *Liverpool Daily News* reported that Cammell Laird's contract to build HMS *Prince of Wales* meant 'spending £5,500,000 across the Empire', and produced a battleship which represented the physical embodiment of the empire's collective strength, the sum of Australian lead, Canadian nickel and timber from Burma and Borneo. Continuing the imperial tradition of the *King Edward VII*-class, the naming of the *Crown Colony*-class cruisers and *Tribal*-class destroyers was highly symbolic at a time when the dependencies were being asked to step up their commitment to the empire's naval defence.[17]

Between 22 February and 6 March 1939, HMS *Ashanti* visited Freetown, Sierra Leone, and Takoradi Harbour in the Gold Coast. While this was intended to 'enable the chiefs and people of the Ashanti to see something of their nameship in the Royal Navy', the tour was filmed to offer those in Britain a glimpse of colonial West Africa. *The Voyage of the Ashanti* continues the imperial propaganda of Admiralty-backed productions like *Britain's Birthright*, visually reinforcing the imperial hierarchy by depicting

Asantehene Prempeh II, the Ashanti ceremonial leader – through whom
Britain indirectly ruled – offering a gold shield and silver bell in homage to
the ship's company. While the Ashanti present good-luck charms to ensure
the ship's survival, its Royal Navy crew are seen placing more faith in their
own technical and professional proficiency through scenes of gunnery
practice, reinforcing imperial dominance through Britain's scientific
rationalism and advancement. The viewer is told that since 'the British
abolished slavery', such principles had developed Kumasi into 'a modern
town, with electric light and telephone', while the fact that 'nothing was
stolen' from the open ship is presented as further evidence of the locals'
moral progress under colonial rule. That West African chiefs and children
alike 'pass along in dumb amazement' at the technological marvel of the
Royal Navy destroyer and view its sailors with 'great curiosity', reinforces
the argument that Britain's paternalistic guidance was still required in that
part of the world. This gap is widened by ethnographic depictions of the
chiefs flanked by war drums and huge ceremonial umbrellas, described to
the viewer as 'a bizarre site, almost magnificent', while the local market
apparently produces an 'indescribable odour', emphasising the region's
otherness and cultural inferiority. Despite such differences, the film tries
to project a reassuring message of imperial strength and unity at a time
when geopolitical tensions and colonial agitation were provoking
insecurity in Britain. The chiefs cooperate in the policy of indirect rule,
being 'enormously proud' of their ship, and when the Gold Coast
Governor, Sir Alan Hudson, sits with them, it is stressed that 'never again
will the Ashanti war drums beat against the British'. As its band perform
beneath the Union Jack, the Gold Coast Regiment are described as 'the
finest native troops in the world'; for them and their naval compatriots,
that would be put to the test barely a month after the film's release, when
Britain declared war on Nazi Germany.[18]


## World War II[xii]

Colonial naval forces across the empire were mobilised following
Germany's invasion of Poland. Members of the Zanzibar Naval Volunteer

---

[xii] See also Duncan Redford, *A History of the Royal Navy: World War II*.

Force undergoing training in Kenya were recalled and sent to sea before war was declared. In Hong Kong, far-removed from the European conflict (though war in China had been ongoing since 1931), members of the Hong Kong RNVR were put in charge of the local flotilla to allow Royal Navy personnel to be redeployed to the Atlantic and Mediterranean theatres. Colonial naval forces not yet established, such as the Trinidad RNVR, were accelerated to fill strategic gaps.

While the major naval threat lay across the North Sea, at least until Japan joined the Axis Powers in December 1941, the Admiralty's strategy was still global and imperial by nature. When during the dark days of 1940, following the fall of France and with the Battle of Britain's outcome still in the balance, plans were drawn up for the Royal Navy to evacuate the British government and royal family to India and Canada and to continue fighting from Halifax. Even though that worst-case scenario never materialised, the naval contribution of Britain's oldest dominion would be crucial to winning the Battle of the Atlantic. With Newfoundland having reverted to being a British colony in 1934 following its economic collapse, the Admiralty requested the Royal Canadian Navy (RCN) base warships in St John's to lead a Newfoundland (later Mid-Ocean) Escort Force. It initially escorted merchantmen to Iceland, a new area under British influence due to its strategic importance and the establishment of the Royal Navy base HMS *Baldur III*, and by February 1942 Canadian warships were accompanying convoys all the way to Londonderry.

By September, around 16,000 Canadians manned 188 vessels across the North Atlantic, almost half of all surface escorts operating in the region. Yet, rapid expansion came at a cost; training had to be curtailed, the RCN's vessels were technologically less advanced and the majority of merchantmen sunk in the theatre occurred on its watch. Canadian warships were therefore redeployed to gain experience along the less-dangerous Britain to Gibraltar route, until March 1943 when the RCN's Rear Admiral Murray inherited the new North-West Atlantic Command at Halifax. In 1944, Canada took sole responsibility for escorting mid-ocean convoys, as its industry churned out 410 merchant vessels, 487 escorts and minesweepers and 3,302 landing craft. While 25 frigates and corvettes, 62 minesweepers and other smaller vessels were supplied to the Royal Navy, the majority of these warships helped grow the Royal Canadian Navy into the world's third largest by the end of the war, having successfully escorted 25,343 merchantmen, 181,643,180 tons of cargo, and North American troops to Europe. Canadian sailors manned the Royal Navy escort carriers HMS *Puncher* and *Nabob* and cruiser HMS *Uganda* (re-commissioned HMCS *Quebec*), RCN destroyers participated in the D-Day landings, while the armed merchant cruiser HMCS *Prince Robert* was part of the British Pacific Fleet task force which reoccupied Hong Kong.[19]

A Canadian also commanded the only colonial naval force in the Caribbean Sea. The Trinidad Royal Naval Volunteer Reserve was formed primarily to defend the largest oil producer in the British Empire, with the island responsible for supplying 38 per cent of its consumption in 1938. That year, the Committee for Imperial Defence, following advice from the Oil Board, ranked the colony higher in strategic importance than Rangoon and Bahrain due to its location west of the Suez–Mediterranean route. Any German occupation would also offer access to the vast oil fields of Venezuela. Furthermore, Trinidad possessed the world's largest asphalt lake, and was strategically placed to protect bauxite shipping from British Guiana, a key component in radio and aircraft production. HMS *Ajax* reported on the island's naval defence in July 1939, and plans to establish a Trinidad RNVR were expedited once war broke out and Port of Spain became the major convoy assembly point in the region due to its location, harbour and resources. As with other colonial naval forces, it was primarily a local patrol and minesweeping unit, and so Lieutenant Commander D.S.G. Lindsay of the RCN minesweeper HMCS *Festubert* was appointed as its first commanding officer.

The Trinidad RNVR's importance increased after Germany deployed 27 U-boats to the Caribbean in 1942 as part of Operation *Neuland*, sinking 385 merchant vessels that year alone. The need for naval expansion to combat the submarine threat came at the same time as Trinidadians were being drawn to higher-paid construction work following the 1940 Destroyers for Bases Agreement. The force already included British, Canadian and South African officers, and Norwegians from the requisitioned minesweeping trawlers *Ornen III* and *Thorvard*. Additional recruits had to be sourced from Britain's other West Indian colonies, with

Fig. 6.5. West Indian ratings of the Trinidad RNVR operating a depth charge
thrower, September 1944

the first volunteers from British Guiana and the Cayman Islands arriving
from July 1941. In addition to manpower contributions, the Caymanian
master shipwright, Captain Rayal Bodden, constructed two wooden
minesweepers for the Royal Navy. By 1945, the Trinidad RNVR had
grown to 75 officers and 1,215 ratings, additionally drawn from
Barbados, Grenada, Tobago, St Vincent, St Lucia, Antigua, St Kitts,
Dominica and Montserrat. While the majority of ratings were classed as
'coloured', only 12 of the officers were non-white.[20]

The Destroyers for Bases Agreement signalled a reassertion of the US
Monroe Doctrine. With repercussions across Britain's Atlantic Empire, an
American military presence was established in Newfoundland, Bermuda,
Bahamas, Jamaica, St Lucia, Trinidad, Antigua and British Guiana. Britain
suspected that the United States would push to take over responsibility for
the South Atlantic, seeing the entire American continent as their sphere of
influence, and that they might outsource the islands' defence to Argentina.
The Royal Navy's presence was essential for upholding Britain's claim in
what remained a significant naval base, despite its decline as a coaling
station. A floatplane base was created on West Falkland in 1939 by HMS
*Ajax* while three other cruisers transported British reservists from

Uruguay for the Falkland Islands Defence Force, and two minesweepers were assigned there the following spring.

Beyond strategic considerations the Falklands also had an imperial importance for Britain's Antarctic claims. It served as a staging post for Royal Navy vessels such as HMS *Carnarvon Castle*, which visited the South Shetland Islands in 1943 to remove any markers supporting Argentine ownership. Lieutenant James Marr of the RNR, a veteran of Ernest Shackleton's expeditions, was reassigned from the Eastern Fleet to lead 14 men in the Colonial Office Operation *Tabarin* during January 1944. His team entered Deception Island's harbour aboard HMS *Fitzroy* and *William Scoresby*, painted over the Argentine flags, and raised the Union Jack, before establishing bases there and on the coast of nearby Graham Island. To ratify Britain's annexation, the Post Office swiftly produced a set of South Shetland stamps bearing King George VI's stately crown atop an icy backdrop. The need to check rival American, Chilean and Argentine interests in the region was also driven by concerns regarding Argentina's pro-Axis sympathies and neutral integrity, which might allow German vessels to use the islands as a base.[21]

On the other side of the Atlantic, Freetown possessed a harbour capable of sheltering 150 ships and it grew to become the major convoy port in West Africa. Like Trinidad, this was accompanied by the establishment of a Sierra Leone branch of the RNVR, with eight officers and 40 African ratings assisting the Royal Navy in keeping the approaches clear, just months after HMS *Ashanti*'s visit. This supplemented the Naval Defence Forces created in the Gold Coast in 1936 (17 officers, 53 ratings), Nigeria in 1939 (55 officers, 600 ratings) and the Gambia Naval Volunteer Force formed in 1938 (8 officers, 63 ratings). Twenty-five thousand African labourers were employed on a huge construction project to transform the Freetown base into a 'young Devonport', equipped with slipways, wireless stations, fuel stores, electricity plants, workshops and a new fresh water supply from the hills. France's fall escalated the danger for merchant vessels passing through the Mediterranean, and those heading east were re-routed around the Cape of Good Hope. This further increased Freetown's importance as 58 ships a day passed through, attracting the attention of German U-boats.

Fig. 6.6. A Royal Navy lifeboat rides 'pick-a-pack' on a motor lorry through
Freetown

Gibraltar also occupied a critical position as the gateway to the
Mediterranean and as the base for Force H, created in June 1940 to fill the
void left by the defeated French fleet before shelling the latter at Mers el
Kebir on 3 July. Force H would also help hunt the *Bismarck*, and escort the
convoys essential for keeping Malta supplied with aircraft and provisions
to repel the German and Italian onslaught. In the process, the Royal Navy
lost a large number of ships near Valletta (21 sunk, plus 13 damaged in six
weeks between March and May 1942).[22]

The Cape route's resurgence increased following the succession of
Japanese victories which pushed the British Eastern Fleet all the way back
to Kilindini harbour in Kenya, and galvanised South Africa's naval
development in Simonstown, Cape Town and Durban. While the South
African Naval Service (SANS) had been reduced to just five officers and
ratings by 1939, 200 members of the Simonstown-based and Admiralty-
administered RNVR(SA) volunteered to serve aboard Armed Merchant
Cruisers, with others employed in examination and signalling duties
ashore. Also recruited on 25 August 1939 was a great dane who

Fig. 6.7. HMS *Queen Elizabeth*, *Resolution*, *London* and *Devonshire* in Valletta harbour

frequented the naval base and enjoyed lying at the top of gangplanks, particularly that of HMS *Neptune*, earning him the name 'Just Nuisance' on his enlistment forms. He sometimes accompanied coastal reconnaissance flights searching for submarines, but his main duty was to improve morale; he broke up fights between human crewmates, though he was personally responsible for the death of HMS *Shropshire* and *Redoubt*'s mascots, and his disrespect for the chain of command meant he was deprived of bones for seven days as punishment for sleeping on a superior's bed!

As a prime architect of the SANS, Field Marshal Smuts was keen to develop the country's naval capability to secure its maritime borders and relieve the Royal Navy's local responsibilities. A new Seaward (aka 'Seaweed') Defence Force (SDF) was formed in November 1940 around an experienced cohort of RNVR(SA) personnel, manning converted whalers requisitioned at the conclusion of whaling season. By the end of 1941, the SDF had grown to 1,232 members operating 24 minesweepers and eight anti-submarine craft. In November that year, Cape Town was visited by HMS *Prince of Wales* and *Repulse* on their way to Singapore, as Churchill hoped Force Z's deployment east would boost empire morale and deter Japan. Large crowds of South Africans came out to witness the battleship and battlecruiser, with 600 cars conveying crewmembers to parties and sightseeing tours as if it were peacetime. On 5 December, just days before the Japanese invasions, *Repulse* showed the flag at Darwin, escorted by the Australian destroyer HMAS *Vampire* in another expression of imperial unity.

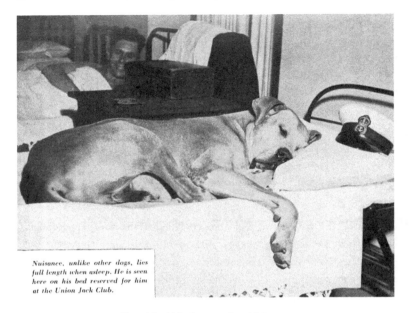

*Nuisance, unlike other dogs, lies*
*full length when asleep. He is seen*
*here on his bed reserved for him*
*at the Union Jack Club.*

Fig. 6.8. Able Seaman Just Nuisance

At the end of World War I, German South-West Africa had been handed to South Africa as a Mandate territory, and this sub-imperial colony also provided an assembly point for east-bound vessels in Walvis Bay, with the SDF providing escorts for convoys heading to Madagascar, Mauritius and the Seychelles. In early 1941, four former whalers comprising the 22nd Anti-Submarine Group were deployed to Alexandria at the Admiralty's behest to serve with the Royal Navy's Mediterranean Fleet. They would be joined by eight South African minesweepers off Libya, while SDF vessels also helped intercept Vichy French convoys near Bellringer and Kedgeree. May 1942's invasion of Madagascar was advocated by Smuts, and undertaken with the South African Rear Admiral Neville Syfret commanding Force H. The following month, British forces also occupied Mayotte, Pamanzi and Dzaoudzi in the French Comoros, while Fortress Diego Suarez remained under British control even after the rest of Madagascar was returned to Free France. By August 1942, the SDF's performance convinced the Admiralty to allow its merger with the RNVR(SA) to form a South African Naval Force (SANF) under dominion control. The Royal Navy supplied three *Loch*-class frigates on permanent

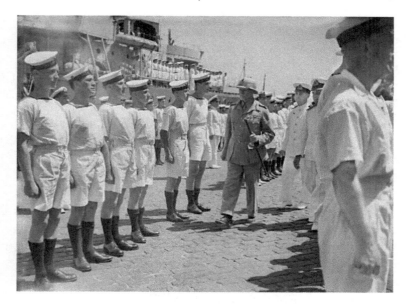

Fig. 6.9. Field Marshal Smuts inspects South African sailors in the Middle East, 16 May 1942

loan, re-commissioned as HMSAS *Good Hope, Natal* and *Transvaal*, and by 1945 the force had grown to 78 escorts, minesweepers and patrol craft, with 9,400 personnel. Around a third of these men were seconded to the Royal Navy, where they served with major warships including the doomed HMS *Hermes* and *Barham,* while South African-built motor launches were also employed by the Royal Navy as far afield as Southeast Asia.[23]

The Kenya RNVR (KRNVR) was the first in the new wave of colonial naval forces, formed in 1933 and co-financed by Uganda and Zanzibar, though suggested a decade earlier with concerns regarding the potential Japanese threat to Mombasa's Kilindini harbour. Sixteen years later, those fears would seem clairvoyant, as Singapore's fall and Japan's bombing of Ceylon forced the Royal Navy's Eastern Fleet to establish its new headquarters at HMS *Tana* in Kilindini. As in Port of Spain and Freetown, a major expansion of the base's facilities followed, employing local labour that, while boosting the local economy, undermined recruitment for the Kenya RNVR who could not compete with the wages offered to construction workers.

From 1938, Zanzibar maintained its own Naval Volunteer Force, with the local Ports and Customs Officer, C.G. Somers, initially commanding three European Officers and 20 'native' ratings. The Police provided musketry lessons, while the officers and six senior ratings trained in Mombasa to help Somers instruct the other ratings in minesweeping, signalling, gunnery and seamanship. The Government vessel *Al Hathera* was converted for use by the force as a minesweeper, with a second smaller vessel, the *Al Nasr*, acquired to act as a second minesweeper and examination vessel. The force's headquarters was in the Cooper's Royal Naval Institute in Mnasi Moja, named after Lieutenant Cooper of HMS *Griffon*, who was killed during a raid on a slaving *dhow* in 1888. The British government levied a fine on the Arabs of Pemba for his murder, and by 1891 had used the proceeds to build a recreation house for sailors visiting Zanzibar. A Tanganyika Naval Volunteer Force also existed between 1939 and 1944, totalling 15 officers and 160 ratings before disbandment, when some personnel moved to the Kenya RNVR.[24]

At the end of 1939, the Kenya RNVR possessed two minesweepers – HMS *Ndovu* and *Nguvu* – and four inland harbour patrol motor launches, operated by 32 European officers, 8 European ratings, four Asian and 44 African seamen. In September 1940, the force was supplemented by the minesweeper *Lindi*, while *Nguvu* was replaced by HMS *Oryx* (renamed *Gemsbuck*). The force played an important role in the Italian Somaliland campaign, where its three minesweepers cleared channels, transported troops and supported raids on Italian positions, while the motor launches, *Baleka*, *Alma* and *Joan* patrolled for *dhows* running the blockade, and signalmen monitored stations at Kismayu and Mogadishu. More recruits were needed for these additional tasks, and the Kenya RNVR grew to 757 personnel by 1 June 1945; this included 654 Africans, though most remained ratings, with only two acquiring the rank of chief petty officer, and 13 petty officers, but no commissioned officers. The Royal Navy's command of the coast and Red Sea helped secure victory in the East African campaign by removing Italy's naval force at Massawa, inhibiting the resupply of its army through blockades, and allowing British forces the freedom of movement to conduct amphibious operations from Mogadishu to Berbera. Military and imperial objectives often dovetailed, with Churchill viewing the capture of Italian Somaliland as important for protecting the route to India, and hoping that Britain might retain its territorial gains well passed the war under some form of a Mandate (British administration would last until 1950).

The Kenyan motor launches *Alma* and *Baleka* were redeployed to Basra following a rebellion in Britain's Mandate territory of Iraq, while HMS *Cockchafer* smuggled the deposed Regent to Trans-Jordan on 2 April 1941. They were joined in the Persian Gulf by HMS *Hermes,* the cruiser HMS *Emerald* and HMS *Impala,* after the former whaler was fitted out in Bombay with a Kenya RNVR crew. The naval presence expanded further ahead of Operation *Countenance* – the invasion of Iran – with the Royal Navy warships, HMS *Falmouth, Snapdragon* and *Shoreham* joined by other empire representatives in His Majesty's Indian Ships *Lilvati* and *Lawrence,* the Australian armed merchant cruiser HMAS *Kanimbla*, HMAS *Yarra* and the RAN-manned *dhow, Naif. Shoreham* began the 25 August offensive, when she fired on the Iranian sloop *Palang* as the task force supported assaults on Bandar-i-Shapur and Abadan by removing other naval threats, providing shore bombardment, landing troops and seizing Axis merchantmen moored there. In October, the pro-British Regent was reinstated at Baghdad, and by the end of the war the Royal Navy occupied bases across the Persian Gulf, at Iran (HMS *Euphrates*), Aden (HMS *Sheba*), Kuwait (HMS *Oman*) and Bahrain (HMS *Jufair*). In India, seven shore bases and depots supported both the Royal Navy and the rapidly-expanding Royal Indian Navy, which from 1944 included a Women's Royal Indian Navy Service (WRINS).[xiii] Indian warships would be deployed as far as the Atlantic and Mediterranean, where they escorted convoys, while HMIS *Jumna* and *Sutlej* also participated in the invasion of Sicily.[25]

Eastern Fleet bases were developed elsewhere in the Indian Ocean, with HMS *Ironclad* in Madagascar, HMS *Sangdragon* in the Seychelles and HMS *Sambur* in Mauritius, where a local Coast Defence Squadron was also established, comprising 12 officers and 108 ratings. Support came in other forms, as the cruiser HMS *Mauritius* was gifted a Walrus sea plane and mess accruements by its colonial public through donations raised during specially orchestrated 'Navy Weeks', a practice replicated elsewhere in the empire. Such sentimental generosity worked both ways; when

---

[xiii]  See also Jo Stanley, *A History of the Royal Navy: Women and the Royal Navy.*

Fig. 6.10. Chief Officer Margaret L. Cooper and Second Officer Kalyani Sen of the
Women's Royal Indian Naval Service, visiting Rosyth on 3 June 1945

cyclones hit the islands in early 1945, causing the death of 24 Mauritians,
the destruction of 10,000 homes, hundreds of fishing boats with
thousands of traps, and a 30 per cent fall in sugar production, a Relief
Fund raised thousands of pounds in donations from HMS *Mauritius*' crew.
The Franco-Mauritian Paul Vigureux also helped the Admiralty Scientific
Service develop an improved system for Royal Navy vessels to detect
underwater echoes from U-boats.[26]

With the Sultan of the Maldives' blessing, the Royal Navy spent
£180,000 for local labour to construct a fleet anchorage at Addu Atoll
beyond the range of land-based aircraft, which became HMS *Maranga*.
Though the Royal Navy would leave within three years, they oversaw a
seismic dislocation of local society. Supposedly positive 'developments'

like new wells, sports fields and cinemas came at huge cost to the natural environment and traditional ways of life. Palm trees were felled, vegetation was concreted over, the sharp influx of motor vehicles caused major congestion, and traditional fishing and farming could not compete with military wages, exacerbating food shortages and resulting in hundreds of chickens, rabbits and goats being introduced from Ceylon alongside alcohol and cigarettes which were religious taboos.

The interwar policy of devolving local patrol and minesweeping duties to colonial reserves fulfilled its intended purpose and reduced pressure on the main fleet, allowing it to concentrate on European waters. This redeployment left peripheral units like the China Station – comprising a few outdated escorts, and *Insect*-class gunboats more suited for traversing the country's vast rivers than fighting a modern naval power, manned by British and Chinese Locally Enlisted Personnel from Hong Kong and Shanghai – looking increasingly vulnerable. Japanese aggression checked the China Station's freedom to operate in the region, and was spurred on by Britain's meek response to 'crises' in Manchuria, Shanghai, Abyssinia and Tientsin, its signing of the Munich Agreement and the withdrawal of British troops from mainland China in August 1940. This left just the gunboat HMS *Petrel* at Shanghai on 6 December 1941, though it refused to surrender and fought bravely, but futilely, against vastly superior Japanese forces including a cruiser, destroyer, gunboat and shore batteries. With the flag of the China Station moved to Singapore along with most of its vessels, only 1,600 Royal Navy personnel, the destroyer HMS *Thracian*, gunboats HMS *Moth* and *Cicala,* eight motor torpedo boats (MTBs), 167 officers and 600 ratings of the local RNVR remained when the Japanese invaded Hong Kong on 8 December 1941.[27]

In the face of overwhelming odds, there was a stoic acceptance that Hong Kong 'could not be held but must be defended'. The Hong Kong RNVR performed many tasks in the MTBs that the vessels were not designed for, including ferrying troops, stores and ammunition, patrols, messenger work, evacuating troops off rocky shores at night 'and anything else that needed to be done'. Their efforts to defend Hong Kong were undermined by the strategic priority given to Singapore, and its Commander in Chief's reallocation of their Lewis guns in exchange for what local officers considered to be an inferior Chinese Bren variety. The problem was exacerbated by the issue of dud ammunition, which produced failure rates of up to 60 per cent, while personnel transfers and the desertion of Chinese T.124 personnel left vessels under strength, meaning they could only man those guns 'more urgent at the moment':

> If *Moth* or *Cicala* or both are supporting the left flank of Army at any time on mainland and are attacked by aircraft whilst bombarding with their 6" armament, I consider that they should temporarily cease fire with A or B Gun and with the 8 men thus released man their pom-pom and 3" until the attack is driven off.[28]

The British forces were 'outnumbered about five to one', and after 'about ten days of non-stop fighting against heavy odds, with hardly any sleep, food or rest and with no reserves at all', the Japanese 'had taken about three quarters of the island' and an order to scuttle was given to prevent the naval vessels from falling into Japanese hands. The Hong Kong RNVR officers of the 2nd MTB Flotilla disobeyed this, however:

> We emphatically refused to scuttle. Around us was the most miserable sight that could greet any seaman's eyes; ships in all stages of being sunk, not by enemy action but by our own hands. It was certainly a black morning. We refused to scuttle because among many reasons we were determined never to give up our boats to the Japs but make a break for it at the last moment.

When word reached the authorities of their planned escape, they were tasked with evacuating China's Chief Admiral, Chan Chak, who had been rallying local Chinese defenders and whose capture, torture or death by the Japanese would cause Britain great embarrassment with the Kuomintang leader Chiang Kai-shek, compromising their wartime alliance. While they succeeded in their dash to the Chinese mainland on Christmas Day, those naval personnel left behind were rounded up in Japanese Prisoner of War camps, where many were interrogated and tortured. Hong Kong had been expected to hold out for 3 months, yet despite the valiant effort of the Royal Navy and its local Chinese and European reservists, the colony was forced to surrender after just 18 days.[29]

In Singapore, a Straits Settlements RNVR was established on 20 April 1934. Instruction initially took place aboard the Governor's yacht, *Sea*

*Belle II*, until 18 January 1935, when the Admiralty provided the *Acacia*-class sloop HMS *Laburnum* on free loan to serve as headquarters and drill ship. The officer in charge, Lieutenant Commander L.A.W. Johnson, had previously been Chief of the Fishery Protection Gunboat, *Liffey*, and was assisted by Chief Petty Officer E.W. Hull, a former instructor at the Royal Navy's torpedo school HMS *Vernon*. The army loaned Sergeant Major Adnan Raji from the Straits Settlements Volunteer Corps to serve as drill instructor, and he became the first Malay Chief Petty Officer. The other officers were Europeans, some with 'experience of nautical matters', though many were 'merchants, brokers, planters and miners'. The ratings were all Malay, some possessing seagoing experience as fishermen or merchant seamen, although others were clerks, *tambies* (office boys) and motor engineers. Training cruises aboard visiting Royal Navy warships and HMS *Terror*, the monitor stationed at Singapore, gave practical experience, but were also intended to 'promote social contacts, and generally to make this unit feel like the navy has taken the RNVR under its wing', encouraging 'esprit de corps and keenness'.

From March 1937, several officers were stationed in Penang, and formed the nucleus of a unit there from October 1938, renamed the Malayan RNVR. It grew from 20 officers and 100 ratings, to 650 ratings and 158 British officers on the eve of World War II. On 4 September 1939, the officers and ratings of both RNVR units were released from civilian employment and mobilised for wartime service. On the same date, the Admiralty and Ministry of War Transport began requisitioning and modifying large numbers of merchant vessels from Singaporean, Malayan and Hong Kong shipping companies, notably the Straits Steamship Company. By 1941, 51 of these had been commissioned into the Malayan RNVR as His Majesty's Ships, armed with cannons, Lewis guns, depth charges and minesweeping gear. A full-time Royal Navy (Malay Section) was also formed in 1939 to relieve Royal Navy personnel for service on other stations. Based in Singapore's HMS *Pelandok* (named for the Malayan mouse deer), the Section was open to Malays aged between 15 and 20, and grew from 400 ratings in 1939 to 1,430 by the end of 1941, with thousands more on the waiting list.[30]

The most infamous naval action involving Singapore was the sinking of HMS *Prince of Wales* and *Repulse* on 10 December 1941, which caused Churchill to proclaim 'I never received a more direct shock [...] the full horror of the news sank in [. . . ] Over all this vast expanse of waters Japan was supreme, and we everywhere were weak and naked'. With the Japanese able to operate with impunity both in sky and on sea, Singapore surrendered on 15 February 1942, the so-called 'worst disaster' and 'largest capitulation in British history'. Although Singapore is widely remembered as an embarrassment for Britain and the Royal Navy, the heroism of Malay sailors was locally celebrated. Like Dunkirk, these men were praised for evacuating Penang and Singapore against the odds, but they paid a high price in the process; only nine of 61 Malayan RNVR ships broke through the enemy cordon to reach Sumatra and Java, the rest were sunk by Japanese forces. Approximately 30 officers escaped, with 50 killed or missing in action, and 49 taken prisoner. For their bravery, they were awarded ten Distinguished Service Crosses, one Distinguished Service Order and eight mentions in dispatches. Of the ratings, 53 were killed, 120 were missing and around 500 were recovered.[31]

Ahmad Dahim bin Noordin joined the Malay Navy as a signalman in October 1941, and, within two months, he found himself at war against Japan. Convinced that British naval power would defend Singapore, he was shaken at seeing the survivors brought in from HMS *Prince of Wales* and *Repulse*: 'We were shocked, because, we read in the papers a few weeks before the outbreak of the war, according to Mr Churchill, British pride of the navy, the *Prince of Wales* and *Repulse*, unsinkable [sic]'.

*Prince of Wales*' nickname of HMS 'Unsinkable' made it more than a ship for Ahmad and his fellow Malay ratings. For them, it was the technological manifestation of the British Empire's professed superiority, and its destruction not only undermined Britain's imperial self-assurance, but also colonial belief in it.

Ahmad did not go to sea – he was engaged in filling defensive sandbags, placing granite rocks on the airfield to prevent enemy landings and guarding military installations. When evacuation was called on 10 February 1942, Lieutenant-Commander H. Vickers, the Malay Navy's Commanding Officer, gave his men the choice of either staying in Singapore or following the British out. Ahmad decided to leave, more from fear than loyalty:

I was really scared because if the Japanese capture us and find out we were former members of a naval unit, maybe they will take us as POWs or maybe kill us. You heard stories of Japanese brutality in China. So I left.[32]

Following Singapore's fall, Ceylon became the most strategically valuable piece of colonial estate in the Indian Ocean. This was where the naval line was drawn, so much that the island's civilian government reported to Admiral Sir Geoffrey Layton following his appointment as Commander in Chief in March 1942. He immediately set about increasing rubber output, formed an Essential Services Labour Corps to counter industrial action, and improved civil defence. Major naval bases were developed at Colombo (HMS *Lanka*) and Trincomalee (HMS *Highflyer*) with the aid of Walker Sons and Company, a local firm of 3,600 engineers usually employed on tea and rubber plantations, who were tasked with repairing and refitting 489 warships and 1,932 merchantmen. Southeast Asia Command (SEAC) also moved its headquarters to Kandy after Admiral Lord Louis Mountbatten was appointed as its Supreme Allied Commander in 1943. This stimulated additional construction projects employing 83,500 people by 1945 and reducing island unemployment to around 1,000, just 4 per cent of its pre-war figure. A Ceylon RNVR also recruited 1,188 officers and men. The influx of capital allowed social welfare policies to be introduced, while their acquiescence to military authority earned local leaders the support of Layton and Mountbatten in advancing the case for self-government after the war.

From Ceylon, HMS *Tactician* secretly inserted Force 136 into Siam. This clandestine group comprised Siamese students recruited from English universities, and it included members of the Malayan RNVR who escaped from Singapore. By August 1945, they had armed several thousand Siamese to fight against the Japanese, while *Tactician* conducted other covert operations, including landing MI6 agents in Penang. Force 136 helped Britain push the Japanese back in Burma, where 139 officers, 1,317 ratings and 9 motor launches of the Burma RNVR landed troops during the Arakan offences. This included the recapture of Rangoon, where the battleship HMS *Queen Elizabeth*, cruisers HMS *Ceylon*, *Suffolk* and *Cumberland*, destroyer HMS *Venus* and carriers HMS *Empress* and *Shah*, provided fire support with the French battleship *Richelieu* and Dutch

Fig. 6.11. Petty Officers Harvey, Hughes and Mason working on a Fairey Seafox at the Royal Naval Air Station at Maharagama, in front of a class of Ceylonese recruits

cruiser *Tromp*. To aid Britain's re-occupation of Malaya, code-named Operation *Zipper*, the Malayan RNVR was reformed with the remaining 300 Malay ratings in Ceylon; the Admiralty hoped that they would help win local 'hearts and minds' and restore confidence in British prestige, which had been so swiftly shattered by the Japanese. Postponed because landing craft were needed for D-Day, the Royal Navy eventually returned on 9 September 1945 to accept the Japanese surrender in Penang with a force of 2 battleships, 4 cruisers, 15 destroyers, 6 escort carriers and 180 naval aircraft.[33]

The 1931 Statute of Westminster had given the dominions the self-determination to decide whether or not to declare war on Germany, and all but the Irish Free State joined Britain in doing so. A consequence of

Fig. 6.12. Men of the Burma RNVR with a Japanese officer's sword they captured

Ireland's decision was the need to form its own naval service, as the Hague Convention required that a neutral country possess the ability to control its own ports and inspect incoming vessels so as to verify their peaceful purposes. Australia and New Zealand were economically dependent upon imperial trade and finance, and merchant sinkings in the Atlantic affected them as much as they did Britain, thus the RAN was placed at the Admiralty's disposal on 30 August, while New Zealand already hosted a division of the Royal Navy. At this point the RAN contained 5,540 personnel, six cruisers, five destroyers and two sloops. The cruisers HMAS *Australia*, *Sydney* and *Hobart*, and destroyers *Voyager*, *Vendetta*, *Vampire*, *Waterhen* and *Stuart* joined Royal Navy forces in the Mediterranean, with *Sydney* sinking the Italian cruiser *Bartolomeo Colleoni* in July 1940 off the coast of Crete. *Perth* joined the East Indies fleet, while the RAN escorted convoys conveying the ANZACs to fight in the Middle East, and its warships supported operations along the North African coast. Loans of personnel and warships between the Royal and Royal Australian Navies continued even after the Japanese offensives saw Australian forces operating under the United States South-West Pacific Area. By the end of the war, Australia's Navy had grown to 40,000 servicemen and 337 ships, while the country provided important logistical support for imperial operations; it surveyed amphibious landing sites, trained 20 per cent of Atlantic anti-submarine personnel in HMAS *Rushcutter*, used its industrial shipyards to repair and refit hundreds of naval and merchant ships, and built 3 *Tribal*-class destroyers, 12 *River* and *Bay*-class frigates and 60 *Bathurst*-class corvettes for the RN and RAN.

New Zealand also acquired its own navy from the war, with the local division of the Royal Navy becoming the Royal New Zealand Navy (RNZN) in October 1941, before taking over Admiralty responsibility for the 30 officers and 476 men of the Fiji RNVR. The New Zealand-manned cruiser HMNZS *Achilles* fought at the battle of River Plate, and a civic reception was held in Auckland for her in February 1940, where Maori representatives presented a kiwi feather *korowai* (cloak) to the cruiser's *rangatira* (leader), Captain William Edward Parry, as a symbol of his great *mana* (authority and prestige). *Achilles'* sister ship HMNZS *Leander* escorted Red Sea convoys, fought in the Mediterranean and hunted raiders in the Indian Ocean, where it destroyed an Italian heavy cruiser. Both cruisers returned to the Pacific to protect New Zealand and Australia from Japanese invasion, and they supported US forces in the bloody battle for the Solomon Islands alongside the minesweepers HMNZS *Matai*, *Tui*, *Kiwi* and *Moa*; the latter two bravely took on the larger, more heavily armed Japanese submarine *I-1* on 29 January 1943, which was wrecked after receiving several rams from *Kiwi* and 4-inch gunfire from *Moa*. The RNZN was augmented by the transfer of British warships such as the cruiser HMS *Gambia*, which then joined the British Pacific Fleet off Okinawa and Japan (accompanied in July and August by the refitted HMNZS *Achilles*). There *Gambia* guarded carriers from *Kamikaze* attacks, while it bombarded the Sakishima airfields and Kamaishi steelworks, becoming the last BFP ship to shoot down a Japanese aircraft, before representing New Zealand at the official surrender ceremony in Tokyo Bay. Seven thousand New Zealanders served with the Royal Navy and provided 10 per cent of the Fleet Air Arm's operational aircrews by 1945;[xiv] they defended Malta, protected convoys in the Mediterranean

---

[xiv] See also Ben Jones, *A History of the Royal Navy: Air Power and British Naval Aviation*.

and Atlantic, attacked the *Bismarck* and *Tirpitz* and helped invade North Africa, Salerno, Sicily and Japan.[34]

For the final months of the war, RAN and RNZN warships served as part of the British Pacific Fleet (BPF), which also included the RCN-manned cruiser HMS *Uganda*. Formed in November 1944, the BPF was based primarily at the newly constructed Captain Cook Dock in Sydney, supported by secondary bases at Cairns, Brisbane, Melbourne, Adelaide and Fremantle, where submarines were deployed from 1942, while the Civil Construction Corps built fleet facilities in New Zealand. The BPF's mission was as much political as it was strategic, intended to reassert Britain's imperial presence and provide a symbol of resurgent naval power in those colonies it had briefly lost and those dominions it had let down, by sending Royal Navy warships to publicly accept Japan's surrender. The British government was especially determined that Rear Admiral Cecil Harcourt reach Hong Kong before the army of Chiang Kai-shek, as even though it fell under the Kuomintang leader's operational sphere, if allowed to fall back into Chinese hands it would be difficult for Britain to enforce the territorial 'rights' it extracted from the Treaties of Nanking and Tientsin. Harcourt entered Victoria harbour on 30 August 1945 ahead of China's Nationalist Army, commanding an awe-inspiring task force led by the battleship HMS *Anson*, the carriers *Vengeance*, *Venerable* and *Indomitable*, the cruisers *Swiftsure*, *Euryalus* and *Black Prince*, four destroyers, eight submarines, six RAN minesweepers and a depot ship. For eight months from 1 September, Harcourt governed the colony as Head of its Military Administration. A smaller task force – featuring the light fleet carrier HMS *Colossus*, the cruisers HMS *Argonaut* and *Bermuda* and four destroyers – occupied Shanghai with the US Navy; HMS *Cleopatra* accepted Japan's surrender at Singapore before Mountbatten's official ceremony, the carrier HMS *Glory* was deployed to Rabaul and the RAN was prominent in retaking the island of Borneo, with Sarawak surrendering to HMAS *Kapunda* on 11 September 1945.[35]

As impressive as these displays of naval theatre were, and as valiantly as its personnel had fought during impossible battles at Shanghai, Hong Kong, and when evacuating Singapore, the Royal Navy could not restore the prestige Britain had lost in failing to safeguard millions of colonial people it had sworn to protect. Furthermore, it had promised to 'respect

Fig. 6.13. D.C. Towell and R. Thompson of 42 Royal Marine Commando with
children from the Tai Po orphanage, during Hong Kong's reoccupation

the right of all peoples to choose the form of government under which
they will live', as enshrined in the Atlantic Charter signed by Churchill
and American President Franklin D. Roosevelt aboard HMS *Prince of Wales*
off Newfoundland in August 1941. While written for European countries
that had fallen under Nazi domination, the statement provided a fillip for
nationalists across the British Empire. Colonial naval service highlighted
the racial inequalities and prejudices inherent in the imperial system,
erupting in protests by members of the Trinidad RNVR in 1943. Worse
would come, as 10,000 members of the Royal Indian Navy mutinied in
February 1946, convincing Prime Minister Clement Attlee that British
rule in the subcontinent was no longer tenable. The following year,
Mountbatten handed power over to the independent states of India and

Pakistan, having overseen their partition as the last Viceroy. Yet this was seen as a setback rather than a sea change for British imperialism, requiring that the Royal Navy change tack to sail the winds of change that would blow across the empire.[36]

CHAPTER 7

# End of Empire?

Swiftly following India and Pakistan in gaining independence was Burma in 1948, a decision derided by Churchill, in naval parlance, as tantamount to a 'scuttle'. Yet the British ship of empire was still very much afloat – it had only altered course. As HMS *Norfolk* conveyed King George VI's representatives, the Duke and Duchess of Gloucester, to Ceylon's independence ceremony, elsewhere Britain pursued a 'second colonial occupation' where its African colonies and Malaya would supply resources for the country's post-war reconstruction.[1]

This was threatened by a Communist insurgency in Malaya, which lasted 12 years after a state of 'Emergency' was declared on 16 June 1948. Under the command of the Penang-born Royal Navy officer, Captain H.E.H. Nicholls, a full-time Malayan Naval Force (MNF) was formed on 24 December 1948, becoming the Royal Malayan Navy (RMN) on 29 August 1952. This operated alongside vessels of the RAN, RNZN and the Royal Navy, which developed a new mode of warfare after forming its first helicopter squadron in 1952 to aid missions against jungle-based guerrillas. The MNF was based in Woodlands Barracks adjoining the Royal Navy's base at Sembawang, Singapore, and was supported by a reconstituted part-time Malayan RNVR. While drawing upon Malay naval veterans from the war, there were attempts to offset the Communist insurgents by indoctrinating more Malayan Chinese ratings into the MNF. Five Harbour Defence Motor Launches, used during the Burma Campaign and left rotting in Keppel harbour, were transferred as patrol vessels and put under the command of two old British reservists and three regular sub-lieutenants seconded from the Royal Navy. Their primary role was to search fishing *kampongs* (villages) and *tongkangs* (boats) for terrorists and weapons smuggled in from Indonesia, assist counter-insurgency operations by transporting Gurkhas up-river, provide fire support, and aid the British 'hearts and minds' campaign by visiting remote villages, opening ships up to the local population and providing entertainment including film screenings.[2]

British influence was preserved in post-colonial countries through Commonwealth economic and strategic ties. In exchange for providing military aid and training to develop Ceylon's armed forces, Britain retained rights to use the naval base at Trincomalee. Partition of the Royal Indian Navy meant that for several years after independence, India and Pakistan continued to rely upon seconded Royal Navy officers to command the senior positions in their new national navies until sufficient local officers were trained to replace them; not until 1958 would Vice Admiral Ram Das Katari take command of the Indian Navy. A paternalistic relationship was thus preserved, inculcating Royal Navy doctrine, sentimental ties and an appreciation of Commonwealth defence strategy, as Indian and Pakistani cadets joined those from the former 'white Dominions', Australia, New Zealand and Canada, in undergoing training at the Britannia Royal Naval College in Dartmouth. They were joined from 1955 by the first Malayan officer cadets, which included Thanabalasingam Karalasingam, the first 'non-Caucasian' to be selected as divisional captain at the BRNC. Thanabalasingam would go on to become the first Malaysian Chief of the RMN on 1 December 1967.[3]

These bonds were reinforced with regular joint exercises involving the Royal and Commonwealth navies, while the latter relied upon the loan or purchase of British warships to build up their fleets and sent representatives to join in the 1953 grand fleet review celebrating Queen Elizabeth II's coronation. An example of these Commonwealth interconnections is HMNZS *Achilles*. Originally loaned by Britain to New Zealand, then sold to India and re-commissioned as INS *Delhi* on 5 July 1948, her former captain, Vice Admiral William Edward Parry, would follow her to become the Indian Navy's new chief a month later. Commonwealth sales of *Colossus* and *Majestic*-class light fleet carriers gave the RAN, RCN and Indian Navy experience of independent carrier operations. Though India opted to become a republic, most

Commonwealth countries retained Britain's monarch as their head of state after independence. A naval presence still accompanied the royal family on official visits, such as in February 1954, when HMS *Ceylon* escorted the royal yacht *Gothic* carrying the young Queen Elizabeth II to Tasmania's 150th anniversary celebrations.

The Royal Navy continued to show the flag in support of Britain's interests, with HMS *Mauritius* steaming 42,976 miles between April 1949 and March 1951 while calling at Gibraltar, Malta, Port Said, Aden, Abadan (where pressure was mounting for Iran to nationalise British Petroleum's oil stakes in the country), Colombo, Bombay, Karachi, Calcutta, Madras, Bahrein, Basra, Kuwait, Mena el Ahmedi, Muscat, Singapore, the Seychelles, Mombasa, Zanzibar, Dar es Salaam, Tanga, Mauritius, Khor Kuwai and Cochin. Royal Navy vessels were essential for assuring regional allies that Britain retained global power and influence, like in 1956 when Gulf sheiks requested the presence of HMS *Superb* in response to local trouble. The ship's marines found that 'the people causing trouble were driving nails through beer bottle tops and spreading them on the road', but 'a march through the neighbourhood by the "Booties" with the band seemed to be effective' in subduing them. *Superb* followed up on this by touring the Gulf and 'entertaining visitors' with unsubtle demonstrations of its firepower. On other occasions, the Royal Navy needed to save British face after failed interventions, such as when HMS *Ceylon* evacuated British troops following an Iraqi coup in July 1958.[4]

At 10:16 on 18 September 1955, Commander Scott of HMS *Vidal* became the last in a long line of Royal Navy officers to formally claim territory for Britain's monarch, when he declared, 'In the name of her Majesty Queen Elizabeth the Second, I hereby take possession of this island of Rockall'. Strategic factors lay behind the decision to annex this tiny inhospitable Atlantic island. Having bought the Corporal Type II nuclear missile from the US, the UK looked to test-launch it from the Uist Isles in the Outer Hebrides. As 'unclaimed territory' 200 miles to the West, it was feared that Rockall 'might well be used by an unfriendly state as a vantage point for the observation of activities upon the range'. The Colonial Office were asked to lay out the correct procedure for augmenting the empire, and after consulting historical precedents, it advised that 'international law probably requires some actual exercise of sovereignty to make the title good [ . . . ] formal annexation could be achieved [ . . . ] by the reading of a proclamation on it by some duly authorised officer'. Prime Minister Sir Anthony Eden and Defence Minister Harold MacMillan decided that a Royal Navy party carrying 'Queen's Instructions' would 'give the act of taking possession of the island as formal and impressive a character as possible'. In the tradition of James Cook's Pacific voyages, colonisation was shrouded beneath scientific guise, with the Admiralty fearing it 'would be severely criticised if it did not make use of the attempt to gather all possible scientific information'; the new survey vessel HMS *Vidal* (named in honour of Rockall's first surveyor) was selected for the mission, with the former Royal Marine James Fisher chosen as 'liaison officer for scientific matters'. Being the first Royal Navy vessel built with a helicopter deck was also important, as this was the safest way to land people on the rock's narrow ledge situated above treacherous surrounding waters. Armed with time-honoured markers of imperial ownership, Commander Scott cemented Britain's claim with a plaque and a flagstaff flying the Union Jack.[5]

A greater test of Britain's imperial status came just a few months later. It withdrew its military presence in the Suez Canal Zone due to rising nationalist agitation, but within days this move emboldened President Gamal Abdel Nasser to nationalise the Canal on 26 July 1956. Constructed at the height of the *Pax Britannica*, the Canal symbolised Britain's imperial and industrial mastery of land and ocean, connecting the Mediterranean and Red Seas so that not even terrain could bar its ships from ruling the waves. Though India was no long the jewel in Britain's empire, Britain still retained significant commitments 'east of Suez', including the counter-insurgency campaign in Malaya, for which loss of the Canal route would carry huge economic, strategic and political ramifications. Consequently, and in collaboration with France and Israel, Britain launched Operation *Musketeer* to regain control of the Canal, bring down the anti-Western Nasser and install a more amenable leader. This bore striking parallels to the invasion of 1882, but Britain was no longer the world's preeminent power. The operation was delayed so that sufficient amphibious forces could be assembled, allowing international opposition to grow, and within 24 hours of the naval offensive on 6

November 1856, Britain had capitulated to American diplomatic and financial pressure, which further diminished its prestige. The tactical advantages of modern helicopters, which had allowed Britain to plant its flag on Rockall, were negated when they deposited Royal Marines on beaches as if they were World War II landing craft, an action described by Marine Major General James Moulton as 'a lash-up of half-forgotten ideas of the Second World War, more apt to an old comrades parade than to a modern war'.[xv] The situation was exacerbated by the fact that British forces were denied use of base facilities in former imperial territories like Jordan and Ceylon due to the political sensitivity and in spite of existing defence agreements. Ceylon's Prime Minister extended this ban to future emergencies, meaning the East Indies Fleet had to be relocated to HMS *Jufair* in Bahrain until the station was disbanded on 7 September 1958.

Suez emphasised the dilemma facing Britain in the era of decolonisation: how to continue protecting its global economic and strategic interests without possessing the colonial territory, military strength or political capital it had in the past. It sought to reduce the Royal Navy's dependency on overseas bases and improve its response to future 'crises', when the Defence White Paper of 1957 announced that the light fleet carrier HMS *Bulwark* would be converted into a helicopter 'commando carrier', a process also carried out on her sister ship, HMS *Albion* in 1962. The landing platform docks (LPDs) HMS *Intrepid* and *Fearless* were ordered alongside six landing ships logistic, while the 1958–9 Naval Estimates retained a carrier battle group permanently on call 'east of Suez', based at Singapore. While the naval dockyards in Hong Kong and Malta were closed and privatised, refuelling stations were maintained at Aden and Mombasa, and a wireless station was retained in Mauritius.[6]

The 1957 White Paper also advocated developing Britain's nuclear arsenal, which was the new measure of 'Great Power' status.[xvi] The Royal New Zealand Navy was enlisted to assist it in Operation *Grapple*, a series of nine British nuclear tests in the Pacific islands of Malden and Christmas between 15 May 1957 and 23 September 1958. The *Loch*-class frigates HMNZS *Rotoiti* and *Pukaki* were deployed to report on the weather

---

[xv] See also Britt Zerbe, *A History of the Royal Navy: The Royal Marines.*

[xvi] See also Philip D. Grove, A *History of the Royal Navy: The Nuclear Age.*

Fig. 7.1. A helicopter and landing craft amphibious assault during the Suez Operation

conditions and detonations, with *Pukaki* positioned closest at 80 nautical miles (148 km) from a three megaton blast (the equivalent of 3 million tons of TNT). Both vessels were present for four of the tests, with *Pukaki* attending an additional five. One of the sailors, Arthur Venus, described the shock wave as 'a dull, thudding noise like a jack-hammer' across the sea, which made 'the bones of his fingers [ . . . ] visible through his eyes'. The ships were wetted down before and scrubbed after the tests, and crew wore anti-flash gear, coveralls, a respirator and dark goggles if nearer to the blast, though no protective gear further away. A 1990 study reported that some of the 500 RNZN personnel present might have contracted leukaemia and haematological cancer as a result of exposure to the radioactivity. Psychological examinations in 2005 showed that they were more prone to depression and possessed poorer mental and physical health due to anxiety, while a 2007 study revealed that nuclear test veterans had also suffered significant genetic damage.[7]

The refitted *Bulwark* was called upon on 1 July 1961, when its helicopters landed Royal Marines of No. 42 Commando into Kuwait to support the pro-British Emir against a suspected Iraqi invasion. By 9 July, Operation *Vantage* had deployed 5,668 British personnel to the country, supported by an offshore fleet containing *Bulwark*, the aircraft carrier HMS *Victorious*, five escorts and an Amphibious Warfare Squadron. Though this projection of naval power achieved its objective in deterring an invasion, in many ways it was illusory, with Britain and the Royal Navy trading in their past prestige. The Middle East Commander in Chief, Air Marshal Sir Charles Elworthy, judged that Kuwait's defences would have fallen by the time the Royal Navy's task force was fully in place, as it had limited fire support until the frigate HMS *Loch Fada* arrived on 5 July, no aircraft carrier until 9 July and no mine countermeasures capacity until 21 July, when the 108th Minesweeping Squadron arrived all the way from Malta. A similar intervention saw the helicopters and Royal Marines of HMS *Albion* quickly quell an Indonesian-instigated uprising by the 'North Borneo Liberation Army' against the Sultan of Brunei in December 1962.[8]

Naval operations abroad were still culturally supported at home. In the same year as Operation *Vantage*, the documentary film *First Left Past Aden* was produced for the Admiralty. It sought to demonstrate to an increasingly ambivalent British public that the Royal Navy still had an important moral mission to carry out in the world, and in the Persian Gulf in particular, where British prestige had been damaged by the events around Suez. The country still retained huge economic interests in the region; it imported two-thirds of its oil from the Gulf, while the sheikdoms conducted business in sterling and invested profits through the City of London, contributing £400 million to Britain's balance of payments. In *First Left Past Aden*, the frigate HMS *Loch Lomond* is seen continuing a long Royal Navy tradition of policing the oceans, bringing peace and stability, and showing the flag of British culture, trade and progress to local peoples who are still framed in anachronistic terms that recall the racial ideology and imperial 'duty' of the Victorian era:

> The Persian Gulf and Arabian seas: what mystic thoughts are conjured in the colourful magic of the name. Caliphs and kasbahs, jewelled scimitars flashing in the sun, tall mysterious minarets and crowded carpet-begging bazaars topple over each other in the confused jumble sale of our imagination [...] The duty has been done: a soothing gesture in the troubled world. The Mullah has been encouraged to express his aims, and in turn will receive the help and understanding of a government at home [ . . . ] Behind this languid air of Jack

there is a sense of purpose running deeply in his veins [ . . . ] [Jack 'Tar':]
'Ruddy important, that's what it is.'[9]

In the film, *Loch Lomond* inhabits a romanticised world where, like the
gunboats of the *Pax Britannica*, Royal Navy officers still possess the power
to influence indigenous rulers and their peoples. Like the civilising
mission of that era, this is not considered a selfish act, carried out for
Britain's own interests, but a moral obligation to paternalistically aid what
are seen as less-'developed' cultures, a role seemingly as timeless as the
Oriental setting it is presented in.

This was despite the wave of decolonisation which was sweeping
through the British Empire during the 1960s, after Prime Minister Harold
MacMillan famously declared that: 'the wind of change is blowing [...]
Whether we like it or not, this growth of national consciousness is a
political fact'. Cyprus, British Somaliland, Nigeria (1960), Sierra Leone,
British Cameroons (1961), Jamaica, Trinidad and Tobago, Western Samoa
(1962), Zambia, Malawi, Malta, Gambia (1964), Botswana, Lesotho,
Barbados, Guyana (1966), Aden (1967), Mauritius, Nauru, Swaziland
(1969), Tonga and Fiji (1970) all gained their political independence over
the course of the decade. In the midst of this, another documentary was
aired in 1966; *Fourteen Hundred Zulu* showed an unfamiliar Royal Navy of
guided missiles and nuclear submarines,[xvii] symbols of modernity and its
Cold War mission with NATO, but a service still committed to protecting
Britain's colonies past and present:

> At this moment, it's 1400 Zulu in Navy time [ . . . ] North of Bermuda, it's
> 10am as a guided missile destroyer heads west nor'west to rendezvous with a
> tanker [ . . . ] 3,000 miles east, near Gibraltar, at 1400, an aircraft carrier
> prepares a division of Buccaneers [ . . . ] A further 5000 miles eastward, it's
> early evening off Singapore where a cruiser's attack radar scans for the echoes
> [ . . . ] in the North Sea, where a coastal mine-sweeper rolls on for her fifth
> sweep of the day over a World War II minefield [ . . . ] It's mid-afternoon
> southeast of Malta, and a commando ship will soon be disembarking vehicles,
> weapons and men [ . . . ] It's 5pm off Aden, where a frigate and her consort are
> hunting [ . . . ] a submerged submarine.

---

xvii  See also Duncan Redford, *A History of the Royal Navy: The Submarine* and Philip D.
Grove, *A History of the Royal Navy: The Nuclear Age.*

The Royal Navy's reluctance to let go of its imperial identity and embrace a more contemporary role drew criticism by the time of the Queen's Silver Jubilee Review. *The Guardian* newspaper reported that the 'tradition that is being nostalgically celebrated at Spithead today is that of Britain's Imperial past', and blamed this 'imperial sentiment [ . . . ] for the limited scale of our submarine programme' and the country's hesitance towards Europe.[10]

While a British Far East Fleet was still maintained at Singapore, the island was decolonised on 31 August 1963 when it federated with British North Borneo, Sarawak and Malaya (independent from 31 August 1957), to form Malaysia. Federation allowed Britain to pass the expense of administering the Borneo territories – which were economically unviable on their own – to Malaya, which needed them to counterbalance Singapore's Chinese population and retain a Malay ethnic majority. It also created a strong Southeast Asian partner to preserve trade, financial and defence connections, which British influence could still operate through. Indonesia's President Sukarno suspected Britain's neo-colonial designs and their effect upon his own regional aspirations, and so launched a policy of *Konfrontasi* (Confrontation), supplying, training and supporting guerrillas in north Borneo to destabilise the new state. A Commonwealth task force was assembled from the Royal Navy, RAN, RNZN and Royal Malaysian Navy to intercept Indonesian raiders, smugglers and infiltrators. Though Malaysia outlived the Confrontation, which was abandoned in August 1966 after Sukarno lost power to Suharto, the federation ejected Singapore for race-based ideological differences, forcing it to become an autonomous city-state that remained heavily dependent on the income and security provided by the Royal Navy base, until industrialisation and national service could be established.[11]

Britain also retained strategic interests in East Africa following the independence of Tanganyika on 9 December 1961, Uganda on 9 October 1962, Zanzibar on 10 December 1963 and Kenya two days after that. The Royal East African Navy, which had grown from the region's colonial naval reserves, was disbanded on 30 June 1962 after Tanganyika withdrew for fear of British interference. Ironically then, the leaders of all four countries were forced to request that their former imperial ruler and the Royal Navy intervene in a series of domestic disturbances which gripped them in January 1964. First, Zanzibar's pro-British Sultan and his Arab-led conservative government were overthrown in a bloody revolution by the socialist African majority on the morning of 12 January. In response, the hydrographic vessel HMS *Owen* was diverted from surveying Kenya's coast to evacuate around 400 British residents, arriving that evening and joined three days later by the frigate HMS *Rhyl* and RFA *Hebe*. It was reported that the evacuees included Sultan Jamshid bin Abdullah, who had briefly served as an Honorary Midshipman aboard HMS *Newfoundland* and *Ceylon* in 1953–4 and ended up exiled in Portsmouth. British action was limited by the speed at which Abeid Karume was swept to power and the country's post-Suez sensitivity to international opinion, meaning that military plans to restore the Sultan were shelved and British influence diminished once its local advisors were replaced by those from Communist countries.[12]

The revolution was swiftly followed by mutinies of the Tanganyika Rifles' 1st Battalion at Colito Barracks outside Dar es Salaam on 19 January, Uganda's Jinja barracks on 23 January and Kenyan soldiers at Lanet armoury the next day. One of the major complaints of the Tanganyika Rifles was the fact they found their promotion prospects blocked by around 50 British officers still employed by the post-colonial government to train local replacements. Such policies were commonly backed by Britain as a means of preserving post-colonial influence, with a similar arrangement having existed in the Indian Navy. While the immediate concern was to protect the lives of British nationals, there was also the spectre of Communist influence spreading across the region, threatening British military bases and staging facilities in Kenya. *Rhyl* was diverted to Tanganyika following the first mutiny, and joining her was the destroyer HMS *Cambrian* and carrier HMS *Centaur*, containing 45 Royal Marine Commando from Aden, by the time the requisite formal requests for intervention were received from Presidents Obote, Kenyatta and Nyerere on 23–4 January. At first light on 26 January, *Cambrian* provided covering fire as Royal Marines were landed on an adjacent sports field and took Colito Barracks with the loss of three Tanganyikans and no British casualties. Mutineers at Tabora and Nachingweya were similarly disarmed and discipline was restored, with the British forces praised in Tanganyika's press and publicly thanked by its government. Yet it would be a hollow victory in that it left Nyerere susceptible to charges of being a 'stooge of the colonialists' from pan-African leaders such as Ghana's President Nkrumah, which pushed him further away from Britain. Kenya welcomed over £1.5 million of defence aid from Britain due to 'the importance attached by the Royal Navy to retaining facilities' at Mombasa, while a team of naval experts was sent to advise on creating a Kenyan Navy. In contrast, Tanganyika replaced the Royal Marines with a Nigerian military presence, sought Chinese assistance to develop a Tanzanian navy in 1966 (having merged with Zanzibar on 26 April 1964), and strongly criticised the limits of British sanctions against Rhodesia during that same year.[13]

Countering British pressure for southern Rhodesia to accept black majority rule, the country's Prime Minister Ian Smith issued a unilateral declaration of independence (UDI) on 11 November 1965. In response, Britain sought a United Nations resolution imposing sanctions, and it fell upon the Royal Navy to enforce a maritime blockade to prevent oil supplies from reaching the wayward colony. A bellicose announcement on 25 February 1966 by Rhodesia's Minister of Commerce and Industry, that 'the day our first tanker arrives in [the Mozambique port of] Beira we shall have won this economic war', prompted the Royal Navy to establish the Beira Patrol. The frigate HMS *Lowestoft* took up position in the Mozambique Channel, to begin interception operations 20–40 miles off Beira on 4 March, and was supported two days later by HMS *Ark Royal*, diverted from the Far East. The carrier's Gannet airborne early warning aircraft extended the search radius to 350 miles beyond Beira, while its Buccaneer strike aircraft and Sea Vixen fighters investigated radar contacts, meaning a tanker could be detected 15 hours before reaching the port. The task force was augmented by a second frigate, HMS *Rhyl*, and a Royal Fleet Auxiliary, while HMS *Eagle* shared carrier responsibilities. Time was an important factor, as the British government had to first seek approval from the flag country before the Royal Navy could stop and board a suspected vessel. This proviso embarrassed Britain on 5 April, when, to great publicity, the *Joanna V* was able to pass into Beira unchecked by the attending HMS *Plymouth*, which had been refused permission to inspect her by the Greek government. Four days later, Security Council Resolution 221 was passed, removing this stipulation and sanctioning the use of 'minimum force'. The reduced requirement for early detection and an agreement with

France meant that the carriers were replaced from 25 May by RAF Shackletons flying from Majunga air base in Madagascar.

By April 1967, two carriers and 17 other warships had served on the station, with two destroyers or frigates remaining on duty. The additional powers granted by Resolution 221 were still insufficient for the Royal Navy to prevent all tankers from breaching the blockade. On 19 December 1967, the French tanker *Artois* rode HMS *Minerva*'s warning shots to reach Mozambican waters, by which time Britain had informed the frigate that *Artois* 'was not carrying oil for Rhodesia'. The Secretary for Defence, Denis Healey, came out in support of the Royal Navy by blaming existing rules of engagement for their 'lack of precision':

> Not only does it place an unfair burden on commanding officers to leave them in any doubt about how far they are expected to go in the enforcement of their requests, but it exposes the Royal Navy to the risk of international discredit should an illegal tanker disregard the threat of force and be allowed to get away with it.[14]

Consequently, naval captains were instructed to fire their main gun at the tanker's engine room, its bridge, or both, until it stopped, and no further attempts were made to run the British blockade.

In spite of the Royal Navy's best efforts, it was estimated in September 1966 that 220,000 gallons of oil per day were leaking into Rhodesia via other ports in the region. The Admiralty calculated that in order to stem this flow by expanding the blockade to all of Mozambique's and South Africa's ports, it would need a non-existent 7 aircraft carriers, and at least 60 frigates or destroyers to maintain 30 constantly on station. Just covering Mozambique's other major port of Lourenço Marques would require an additional 15 escorts, 4 RFAs and £1.5 million per month, seriously undermining the Royal Navy's other global commitments for an outcome which would 'not really bite' because the South African route remained open. Britain could expect no favours from its former dominion, with relations becoming increasingly strained under the Afrikaner-led National Party. The Simon's Town Agreement passed control of the naval base to the South African Navy in 1957, and while Britain retained the right to use its facilities in exchange for warship sales amounting to four frigates, ten minesweepers and five seaward defence vessels, South Africa had become a republic and left the Commonwealth in 1961. International condemnation of the country's Apartheid policy brought UN resolutions restricting arms sales to South Africa, and Britain's 1964 Labour government refused to accept additional naval orders. The political impact of this upon Simon's Town was lessened by defence cuts, which ended the Royal Navy's permanent presence in the South Atlantic and withdrew the frigate HMS *Lynx* from the station on 17 June 1967, with the agreement formally terminated on 16 June 1975.[15]

While the Royal Navy carried out the Beira Patrol with professionalism, service morale nevertheless suffered due to the frustrating futility of its efforts. It could not be abandoned due to Foreign and Commonwealth Office concerns about the political message it would send out, providing a fillip to the Smith government as 'proof' of Britain's impotence, while undermining the country's claim that it remained a major global player. While Britain appeared to have learned the new rules of the game since Suez, having secured international backing, it then acted unilaterally, viewing Rhodesia's rebellion as an imperial matter when a coalition navy called under paragraph five of Resolution 221 might have increased the blockade's effectiveness but fallen under foreign direction. This was somewhat ironic considering that the CVA-01 replacement aircraft carrier programme was also axed in 1966 following major financial pressures, theoretically signalling an end to the Royal Navy's independent, expeditionary role.[xviii] In spite, and perhaps because, of decolonisation, Britain especially did not wish to lose face in the eyes of newly independent African and Asian members of the Commonwealth, who pushed for firmer action against Rhodesia. In many parts of the world, the presence of a Royal Navy warship still projected an impression of power and prestige, the psychological reason why negotiations with Smith were held aboard the cruiser HMS *Tiger* in December 1966 and LPD HMS *Fearless* in October 1968. In the Foreign and Commonwealth Office's opinion, 'the potential savings to our resources accruing from foreign participation in the patrol' were insufficient 'to justify the substantial risk inherent in such an initiative of embarrassment at the UN and of a rebuff from the governments we approach'.

---

[xviii] See also Ben Jones, *A History of the Royal Navy: Air Power and British Naval Aviation*.

Hot on the heels of CVA-01's cancellation was the 1967 announcement by Harold Wilson's Labour government that Britain would close its remaining bases 'east of Suez' following a phased withdrawal of its forces. The wheels had been set in motion by the time the Conservatives' Edward Heath replaced Wilson as prime minister in June 1970; that year, the Royal Navy closed its final Persian Gulf base in Bahrain, and the last Commander in Chief of the Far East station lowered his flag at Singapore, though Royal Navy warships retained a reduced commitment to Southeast Asia as part of the UK's Five Power Defence Agreement with Australia, New Zealand, Malaysia and Singapore. Other bases in the region were paid off by 1976, including HMS *Mauritius*, after those same islands had been suggested as a potential site in the RAF's 'island hopping' strategy, which had helped kill the replacement carriers. It was not all withdrawal, however; in 1969, two frigates were sent with a detachment of metropolitan police constables to reoccupy the Caribbean island of Anguilla after it refused to federate with St Kitts and Nevis. While Wilson wanted to prevent the island from becoming a regional base for drug smuggling, the 'bloodless invasion' was portrayed as something of an old-fashioned farce by British newspapers.[16]

The Beira Patrol was reduced to a single frigate from 1971 until 25 June 1975, when HMS *Salisbury* brought the operation to a close after Mozambique gained independence from Portugal and guaranteed to halt the passage of oil. Despite the Royal Navy's best efforts, the Beira Patrol's effectiveness was undermined by the operational limitations imposed on it by a British government more concerned with allaying international criticism and preventing a military escalation than forcing Smith's capitulation. At the height of operations between March 1966 and March 1971, 47 tankers were intercepted, of which 42 were allowed to proceed to Beira while the remaining five either refused to stop or were escorted elsewhere. In contrast, over just 13 months between April 1966 and May 1967, 169 tankers entered Lourenço Marques while oil was also supplied via South Africa, though it cost Rhodesia more to transport oil via that route than through Beira's pipeline.[17]

Perhaps an unexpected dividend from this international 'face saving' exercise came in 1982, when African Commonwealth countries which had called for action against Rhodesia, like Kenya and Uganda, then supported Britain in the United Nations when Argentina captured the Falkland Islands. More of the empire had been decolonised, including the Bahamas (1973), Grenada (1974), the Solomon Islands, Ellice Island and Dominica (1978), the Gilbert Islands, St Lucia and St Vincent (1979), Antigua and British Honduras (1981), while Britain had removed its military personnel from Malta in 1979, and Rhodesia became officially independent – as Zimbabwe – in 1980. The withdrawal of the Royal Navy's Antarctic Patrol Ship HMS *Endurance* in 1981 seemingly verified Britain's imperial retrenchment, encouraging Argentina's military junta to invade the Falkland Islands on 2 April 1982. Instead, Britain embarked upon another unilateral campaign against an illegitimate change in colonial government, spearheaded by the Royal Navy. New Zealand offered naval assistance, but the distance and logistics involved were prohibitive to swift action, while Canada's Prime Minister, Pierre Trudeau, favoured a policy of neutralism.[18]

The Senior Service was in a very different state compared to 1966. Nowhere was this more apparent than in the Fleet Air Arm. HMS *Victorious* had been sold in 1969, while of the two carriers which led the early months of the Beira Patrol, *Eagle* was broken up and *Ark Royal* was decommissioned in 1978, taking their fixed-wing aircraft with them. With CVA-01's cancellation, virtually all that remained were HMS *Hermes*, the last converted 'commando carrier', the 'through-deck cruiser' HMS *Invincible*, designed for anti-submarine warfare and due to be sold to Australia, and the experimental sub-sonic harrier 'jump jet'. Despite the limitations, First Sea Lord Sir Henry Leach convinced the British Cabinet that a task force could be assembled within 72 hours to retake the islands, though he was under pressure to prove the Royal Navy's worth after 'The Way Forward' White Paper of June 1981 advocated even deeper cuts to the service. Without Leach and the Royal Navy's advocacy, the operation might never have happened, as both the army and RAF considered it hopeless without substantial air cover, an assessment shared by the US Navy who dismissed Britain's chances as a 'military impossibility'. For many outside observers, Britain's decision to re-'take' the Falklands represented a continuation of its imperial tradition. This was encapsulated by the front page of the American magazine, *Newsweek*, which borrowed its 19 April headline from the popular *Star*

*Wars* sequel, *The Empire Strikes Back*, above a picture of *Hermes* bedecked
with harriers and helicopters as she steamed to the South Atlantic.[19]

Virtually the entire Royal Navy, as it remained, was mobilised for the
operation; 51 warships, plus 45 merchant vessels and cruise ships
requisitioned to transport troops, ammunition and supplies for the task
force. On 26 April they recaptured South Georgia, before landing on the
Falklands on 21 May. After Argentina surrendered on 14 June, Prime
Minister Margaret Thatcher also drew parallels with Britain's imperial
past in celebrating the rebirth of 'that spirit which has fired her for
generations past', upon which the country's greatness depended:

> We have ceased to be a nation in retreat. We have instead a newfound
> confidence – born in the economic battles at home and tested 8000 miles away
> [ . . . ] we can rejoice at our success in the Falklands and take pride in the
> achievement of the men and women of our task force. But we do so, not as at
> some flickering of a flame which must soon be dead. No – we rejoice that
> Britain has rekindled that spirit which has fired her for generations past and
> which today has begun to burn as brightly as before. Britain found herself again
> in the South Atlantic and will not look back from the victory she has won.[20]

Nineteenth-century ultra-imperialists had similarly believed that Britain
would 'irresistibly fall into national sluggishness of thought, were it not
for the worldwide interests given us by the necessity of governing and
educating the inhabitants of so vast an empire'. Though most of the
formal empire had gone, a renewed doctrine of Palmerstonian liberal
interventionism offered Britain an opportunity to regain a leading role in
the world and sense of national purpose – albeit often as part of a
coalition. The policies of the 1981 White Paper were revised, the Royal
Navy retained its amphibious warships and the decision to sell HMS
*Invincible* was reversed, though this had the knock-on effect of ending the
RAN's own carrier capability, following HMAS *Melbourne*'s decom-
missioning in May 1982. The defences of the Falkland Islands were also
strengthened, with the construction of a new airfield at Port Stanley and a
permanent garrison including a Royal Navy patrol vessel. In 1991, Britain
intervened in Kuwait as part of Operation *Desert Storm*, where it had
longstanding interests and had sent the Royal Navy to defend against Iraqi
aggression 30 years previously. This time the Royal Navy deployed three

7.2 HMS *Invincible* returns to Portsmouth amid massive celebrations, 17 September 1982

destroyers, four frigates, three mine-hunters, two submarines and helicopters from five Fleet Air Arm squadrons, which accounted for the destruction of 15 Iraqi vessels while Royal Marines fought ashore. Britain would follow this with several other 'liberal interventions' over the next two decades, in Bosnia (1992–6), Iraq (1998), Kosovo (1999), Sierra Leone (2000), Afghanistan (2001–14), Iraq again (2003–9) and Libya (2011). The Royal Navy continues to be crucial for Britain's ability to conduct this type of expeditionary campaign, a role which looks far from over with the first of two 65,000 ton *Queen Elizabeth*-class aircraft carriers, the largest warships ever built in Britain, due to be commissioned in 2017.[21]

Yet this period also witnessed the latest (though perhaps not last) case of British decolonisation, when Hong Kong was handed back to China at the expiration of the New Territories' 99-year lease on 30 June 1997. As early as the 1920s, Britain admitted that it could not successfully defend the colony, a fear that manifested itself with Japan's swift invasion. China's rise only strengthened this reality, and meant that for most of the post-

war period Britain tried to avoid a confrontation over Hong Kong; when
the sloop HMS *Amethyst* was fired upon by Chinese Communists in 1949
after being grounded in the Yangtze river, Britain's only response was to
free the ship and escape, a far cry from the Opium Wars. As Hong Kong
Island and Kowloon could no longer be disentangled from their northern
hinterland, it was politically and economically more beneficial for Britain
to earn Chinese favour by handing the entire colony back. This overrode
concerns about China's human rights record and the prospects it held for
Hongkongers, who had been fed 'British values' such as liberalism,
democracy and freedom of speech, the consequences of which reverberate
in political protests, as witnessed in 2014. Just as the Royal Navy had been
pivotal to the birth of the colony, so it occupied a prominent place at the
end; after lowering the White Ensign at HMS *Tamar*, the frigate HMS
*Chatham* escorted the royal yacht *Britannia*, carrying a teary Chris Patten,
the last Governor, off into the night as the sun finally set on Britain's
Empire in the East.[22]

Yet 14 British Overseas Territories remain: Akrotiri and Dhekelia (in
Cyprus), Anguilla, Bermuda, British Antarctic Territory, British Indian
Ocean Territory, the British Virgin Islands, the Cayman Islands, the
Falkland Islands, Gibraltar, Montserrat, the Pitcairn Islands, Saint Helena
Ascension and Tristan da Cunha, South Georgia and the South Sandwich
Islands, and the Turks and Caicos Islands. Some do so because they would
be economically unsustainable as independent states, such as the remote
Pacific island of Pitcairn, which possesses a population of just 48 people,
mostly descended from *Bounty* mutineers and their Tahitian wives. Others
have enjoyed a huge rise economically and in their standard of living, such
as the Cayman Islands, whose growth as an international financial centre
and tax haven was kick-started by remittances sent home by Caymanians
who served in the Trinidad RNVR during World War II and subsequently
gained employment with international shipping companies due to the
professional experience they acquired from the Royal Navy.[23]

The sovereignty of Britain's Overseas Territories continues to be
contested, despite the politically sensitive rebranding of its remaining
'colonies'. They represent ten of the 16 Non-Self-Governing Territories
listed by the United Nations' Special Committee on Decolonisation,
though the British government has called for them to be 'de-listed' after a

June 2012 White Paper determined that they already possessed a measure of self-government. With a prevailing air of paternalism, it argued that the UK had a fundamental responsibility to ensure the Territories' security, good governance and respect for British values regarding human rights and the rule of law, but, crucially, it was for the people of the Territories to decide whether to maintain the relationship.[24]

One of the disputed Territories, Gibraltar, on 4 August 2004 celebrated the tercentenary of its conquest by the Royal Navy's Admiral Sir George Rooke. Almost the entire population of 'the Rock' linked hands as they lined the Territory's borders, a symbol of their unity with one another and Britain, which physically excluded Spain on the other side. The First Sea Lord was present with the Defence Secretary when the frigate HMS *Grafton* fired a 21-gun salute as part of the festivities. In response, the Madrid newspaper *El Pais* was told by Spain's Foreign Minister that he found it 'very strange that in this twenty-first century, the military occupation of part of an EU member-state's territory is commemorated by another member state'. While critics deride Gibraltar's strategic significance to present-day Britain as 'rather less than Rockall' and its retention as a 'saloon-bar folly' in foreign policy terms, they fail to consider the role of British identity for Gibraltarians, and the residents of Britain's other Overseas Territories.[25]

Referendums were held on 7 November 2002 and 10–11 March 2013, when 98.45 per cent of Gibraltarians said 'no' to a power-sharing agreement with Spain and 99.8 per cent of Falkland Islanders voted to remain a British Overseas Territory, though Spain and Argentina, as rival claimants – based on their own colonial history – naturally dispute the legitimacy of the polls. In the build-up to the Falklands vote, Argentina's President Cristina Kirchner argued that Britain 'continues to be a crude colonial power in decline', and took to social media to accuse the UK of practicing 'typical nineteenth century colonialism. [And an] Anachronistic use of force in violation of international law'. Prime Minister David Cameron's response was that Britain 'supports the Falkland Islanders' right to self-determination, and what the Argentines have been saying, I would argue is actually far more like colonialism because these people want to remain British and the Argentines want them to do something else'. It is this right to self-determination and a more culturally inclusive, less racially defined concept of 'British' identity that characterises the UK's contemporary relationship with its last Overseas Territories; they remain defended by its warships, while the colonies of the past were often forced to accept Britain's rule under the barrel of a Royal Navy gun.[26]

# Conclusion

Britain's imperial and maritime histories are intrinsically interwoven into the fabric of the nation. Were it not for the existence of the Royal Navy, there could not have been a British Empire, and without an empire the navy would not have been what it was: the most powerful fleet in the world, whose influence stretched beyond the seas and imperial borders to shape the lives of billions of people across generations. The hard and soft power exerted by both British imperialism and the Royal Navy was invariably interconnected.

Colonialism provided the navy with the building blocks for a strong fleet, supplying natural resources and markets which stimulated inter-continental trade. This produced revenues which were reinvested to meet demands for better naval protection of that commerce, while the expansion of mercantile shipping increased the quality, quantity and experience of British seamen who could reinforce the Royal Navy's ranks during wartime. The improved quality of naval sailors and ships, reinforced and repaired by colonial manpower in safe overseas harbours, provided a platform for victories which expanded Britain's territorial holdings and strengthened this cycle. More resources, markets and a larger, battle-hardened navy suppressing piracy and slavery increased flows of trade, which accelerated Britain's industrialisation ahead of its rivals. This provided the Royal Navy with a technological advantage in iron-clad armour and steam-powered speed and manoeuvrability, which gave it greater diplomatic weight and opened up new regions for British commercial exploitation. Material strength provided prestige and psychological power, which allowed Britain to acquire influence 'on the cheap' in areas beyond its formal rule.

Yet this increase in contact, compromise and Westernisation undermined the authority of collaborative rulers and brought social and political instability, which threatened British economic investments. The Royal Navy was called upon to restore a favourable equilibrium, quelling opposition in the short-term, but often forced to establish more direct British control through territorial annexation that extended the borders of the empire. Naval officers appointed to key positions governing the colonial administrations, assisted by visiting warships, helped strengthen Britain's imperial power by cultivating a conducive environment for its merchants, missionaries and settlers to operate in, spreading 'British values' such as Christianity, free trade and the rule of law. The economic development of the colonies and the inculcation of local populations created additional resources and patriotism for mobilisation when imperial rivalries exploded into world warfare. While the empire contributed men, ships, money and bases that helped Britain win, imperial hierarchy and the strategic prioritisation of the 'motherland's' defence over colonial 'cubs' she had sworn to protect, fuelled nationalism and desires for both political and naval self-determination.

Constant imperial expansion was unsustainable, eventually meaning Britain could no longer finance a Royal Navy large enough to police all of its global interests, and it had to entrust an increasing devolution of the empire's naval defence to its colonies and dominions. The Royal Navy still needed to second officers, loan and sell vessels, and provide training and expertise to develop Commonwealth navies and their professional associations which Britain hoped would preserve its influence once formal imperial power reverted to more informal decolonised connections. Warship visits sustained defence agreements and fleet interoperability through joint exercises, while favourable sentiment allowed Britain to retain base rights and procurement contracts in post-colonial countries, and secure international support when its residual interests were threatened.

Though some symbols of colonial ownership were removed when the Union Jack was lowered at independence, many markers of the Royal Navy's imperial influence remain, personified for perpetuity through names and memorials. Naval explorers personally branded their 'discoveries' and patented them with charts, establishing a view of the world still recognised in the digital maps of today. Both the natural environment of islands (Admiralty, Carteret, Chatham, etc), mountains (Erebus), straits (Dampier), and harbours (Keppel), and man-made scenes of cities (Vancouver), roads (Captain Cook Highway), canals (Dickens), universities (Flinders), statues, public houses and even countries (the Cook Islands), constructed by communities of colonists following the paths the navy plotted overseas, echo with the names and exploits of its officers and ships.

The Royal Navy affected cultural as well as physical landscapes. It imported clothes and practices like tattooing from indigenous societies it came into contact with, which then became fashionable in Britain, while tales of its encounters with the exotic inspired romanticism and popular entertainment, becoming the subject of literature, art and theatre. It had a similar effect on the development of colonial cultures; for the colonisers, it preserved their sense of Britishness and connection to home by enabling and entertaining an elite world of parties, balls and banquets; for the colonised, traditional customs became merged with naval imports such as uniforms, bands and parades, to produce new indigenous forms of dance and music. With the imperial spread of the English language becoming a tool for subjugation, naval words and phrases entered popular usage (for example, in Fiji the local Kava drink became known as 'grog'). While the Royal Navy traversed both sides of colonial society through sporting competitions, concerts and public displays, it still ascribed roles according to class and race. As establishers of first contact, naval representations of 'native' people in drawings, letters and memoirs informed a racialised way of interpreting and reordering the world to justify one group's subjugation by another . This fostered negative stereotypes, prejudice and racism, which influenced the Royal Navy's recruitment and management of colonial personnel, and haunts society still.

Royal Navy vessels were thus carriers of intellectual as well as biological contagion. While countless people died as a result of naval gunfire, and others became infected with diseases transmitted by its sailors, more lost their lives from the violence and upheaval caused by territorial dispossession and the dislocation of indigenous belief systems and political structures, ironically as 'civilised' behaviour concerning a lawful respect for life and property was being asserted. Royal Navy warships defended slavers until it became economically more beneficial for Britain to suppress them in favour of 'legitimate' trade; they removed animals from their natural habitat to feed or entertain their crews as 'mascots', while wildlife was destroyed to construct naval bases, unearth fuel and acquire the natural resources required to build and maintain a fleet.

Yet it should not be forgotten that the Royal Navy was first and foremost a national navy and while its actions, like those of the empire, could sometimes be framed as either altruistic or cruel, they were ultimately carried out for Britain's overriding benefit. In the process, its influence was felt on every sea and continent, forever changing them for better or worse, and making the Royal Navy a fundamental force in shaping the world as we know it.

# Notes

## Introduction

1 Ngũgĩ wa Thiong'o, *Decolonising The Mind: The Politics of Language in African Literature* (London, 1986).

2 Jan, Rüger, *The Great Naval Game: Britain and Germany in the Age of Empire* (Cambridge, 2007).

## Chapter 1. An Empire Emerges

1 Rear Admiral Chris Parry, 'Britain has to decide upon the Royal Navy's role', *The Telegraph*, 19 March 2012.

2 Herbert A.L. Fisher, *The History of England from the Accession of Henry VII to the Death of Henry VIII, 1485–1547* (London, 1910), p. 168.

3 William H. Sherman, 'Putting the British Seas on the map: John Dee's imperial cartography', *Cartographica* XXXV/3/4 (autumn/winter, 1998), pp. 1–10; Graham Pechey, ' "Empire laid up in heaven": Postcoloniality and eternity', *Critical Quarterly* XLIX/2 (summer, 2007), pp. 1–30.

4 Pechey, 'Empire laid up in heaven', pp. 3–4.

5 Lesley B. Cormack, *Charting an Empire: Geography at the English Universities 1580–1620* (Chicago, 1997), pp. 1–5; Lesley B. Cormack, 'Britannia rules the waves?: Images of empire in Elizabethan England', in Andrew Gordon and Bernhard Klein (eds), *Literature, Mapping and the Politics of Space in Early Modern Britain* (Cambridge, 2001), pp. 45–68, 53.

6 Daniel A. Baugh, 'Great Britain's "blue-water" policy, 1689–1815', *The International History Review* x/1 (February, 1988), pp. 33–58; Francisco J. Borgie, 'Richard Hakluyt, promoter of the New World: The navigational origins of the English nation', *Sederi* xIII (2003), pp. 1–9.

7   David A. Boruchoff, 'Piety, patriotism, and empire: Lessons for England, Spain, and the New World in the works of Richard Hakluyt', *Renaissance Quarterly* LII/ 3 (fall, 2009), pp. 809–58; Borgie, 'Richard Hakluyt, promoter of the New World', pp. 1–9, 2–6.

8   Cormack, 'Britannia rules the waves?', pp. 45–68; Richard Deacon, *John Dee* (London, 1968), pp. 92–5; N.A.M. Rodger, 'The Drake-Norris expedition: English naval strategy in the sixteenth century', *Militaria: Revista de Cultura Militar* VIII (1996), pp. 89–98.

9   David Harris Sacks 'The true temper of empire: Dominion, friendship and exchange in the English Atlantic, *c.* 1575–1625', *Renaissance Studies* XXVI/4 (September, 2012), pp. 531–58, 532–4.

10  Elaine Murphy, 'The navy and the Cromwellian conquest of Ireland, 1649–53', *Journal of Maritime Research* XIV/1 (2012), pp. 1–13.

11  Paul Kennedy, *The Rise and Fall of British Naval Mastery* (London, 1976), pp. 55–7; John Gallagher and Ronald Robinson, 'The imperialism of free trade', *Economic History Review* VI/1 (1953), pp. 1–15.

12  Edmund Malone (ed.), *The Critical and Miscellaneous Prose Works of John Dryden*, Vol. 1 Part 2 (London, 1800); Crista Knellwolf, 'The exotic frontier of the imperial imagination', *Eighteenth-Century Life*, XXVI/3 (fall, 2002), pp. 10–30.

13  Kennedy, *The Rise and Fall of British Naval Mastery*, p. 64.

14  John Evelyn, *Navigation and Commerce, the Origins and Progress* (London, 1674), pp. 15–17, 32–3; Steven C.A. Pincus, 'Popery, trade and universal monarchy: The ideological context of the outbreak of the second Anglo-Dutch War', *The English Historical Review* CVII/422 (January, 1992), pp. 1–29.

15  Daniel A. Baugh, 'Maritime strength and Atlantic commerce: The uses of "a grand marine empire"', in Lawrence Stone (ed.), *An Imperial State at War: Britain From 1689 to 1815* (London, 1994), pp. 185–203.

16  To Governor Fletcher, February 1697, J.W. Fortescue (ed.), *Calendar of State Papers: Colonial Series, America and West Indies, 1696–97* (London, 1904), p. 343; Gerald S. Graham, 'The naval defence of British North America 1739–1763', *Transactions of the Royal Historical Society* XXX (December, 1948), pp. 95–110.

17  Brian Lavery, *Shield of Empire: The Royal Navy and Scotland* (Edinburgh, 2007), pp. 23–4; J. Grant (ed.), 'The old Scots navy from 1689 to 1710', *The Navy Records Society*, Vol. 44 (London, 1914), p. 48.

18  Malachy Postlethwayt, *The Universal Dictionary of Trade and Commerce, with Large Additions and Improvements*, Vol. 2 (London, 1774); Baugh, 'Maritime strength and Atlantic commerce', pp. 185–203.

19  Paul Langford, *Short Oxford History of The British Isles: The Eighteenth Century:1688–1815* (Oxford, 2002), p. 215; Charles Elkins, *The Naval Battles of Great Britain: From the Accession of the Illustrious House of Hanover to the Throne to the Battle of Navarin* (London, 1828), p. xvi–vii; Baugh, 'Maritime Strength and Atlantic Commerce', pp. 185–203.

20    Baugh, 'Maritime Strength and Atlantic Commerce', pp. 185–203; Gallagher and Robinson, 'The imperialism of free trade', pp. 1–15.

21    Alfred Thayer Mahan, *The Influence of Sea Power Upon History, 1660–1783*, 12th Edition (Boston, 1890/1918), pp. 328–9; Kennedy, *The Rise and Fall of British Naval Mastery*, p. 107.

22    Michael Duffy, 'The establishment of the Western Squadron as the linchpin of British Naval strategy', in M. Duffy (ed.), *Parameters of British Naval Power* (Exeter, 1992), p. 75; Graham, 'The naval defence of British North America 1739–1763', p. 108.

23    T.D. Potgieter, 'Maritime defence and the South African Navy to the cancellation of the Simon's Town Agreement', *Scientia Militaria: South African Journal of Military Studies* XXX/2 (2000), pp. 159–82.

24    Charles R. Foy, 'Uncovering hidden lives', *Common-Place* IX/2 (January, 2009), http://www.common-place.org/vol-09/no-02/tales/ (accessed 11 December 2013); N.A.M. Rodger, *The Wooden World: An Anatomy of the Georgian Navy* (New York, 1986), pp. 159–61.

25    John N. Grant, 'Black immigrants into Nova Scotia, 1776–1815', *The Journal of Negro History* LVIII/3 (July, 1973), pp. 253–70.

26    Frank A. Cassell, 'Slaves of the Chesapeake Bay area and the war of 1812', *The Journal of Negro History* LVII/2 (April, 1972), pp. 144–55.

27    Nathaniel Millett, 'Britain's 1814 occupation of Pensacola and America's response: An episode of the war of 1812 in the southeastern borderlands', *The Florida Historical Quarterly* LXXXIV/2 (fall, 2005), pp. 229–55.

28    Rodger, *The Wooden World*, pp. 159–61; Millett, 'Britain's 1814 occupation of Pensacola and America's response', pp. 238–9.

29    Grant, 'Black Immigrants into Nova Scotia', p. 268.

## Chapter 2. Science and Exploration

1     Glyndwr Williams, ' "To make discoveries of countries hitherto unknown", The Admiralty and pacific exploration in the eighteenth century', in Alan Frost and Jane Samson (eds), *Pacific Empires: Essays in Honour of Glyndwr Williams* (Melbourne, 1999), pp. 13–14.

2     William Dampier, *A Voyage to New Holland, Etc., In the Year 1699* (London, 1729); Geraldine Barnes and Adrian Mitchell, 'Measuring the marvelous: science and the exotic in William Dampier', *Eighteenth-Century Life* XXVI /3 (fall, 2002), pp. 45–57.

3     Kapil Raj, '18th-century pacific voyages of discovery, "big science", and the shaping of an European scientific and technological culture', *History and Technology: An International Journal* XVII/2 (2000), pp. 79–98.

4   Williams, 'To make discoveries of countries hitherto unknown', p. 18; Raj, '18th-century pacific voyages of discovery', pp. 79–98.

5   'Documenting a Democracy', *Museum of Australian Democracy*, New South Wales transcripts,    http://foundingdocs.gov.au/resources/transcripts/nsw1_doc_1768.pdf (accessed 21 December 2014); Raj, '18th-century pacific voyages of discovery', pp. 79–98.

6   Daniel Clayton, 'The creation of imperial space in the Pacific Northwest', *Journal of Historical Geography* XXVI/3 (2000), pp. 327–50; Richard Sorrenson, 'The ship as a scientific instrument in the eighteenth century', *Osiris*, 2nd Series XI (1996), pp. 221–36; Raj, '18th-Century pacific voyages of discovery', pp. 79–98.

7   Sorrenson, 'The ship as a scientific instrument in the eighteenth century', pp. 221–36.

8   Raj, '18th-Century pacific voyages of discovery', pp. 79–98; Janet Owen, 'Collecting artefacts, acquiring empire: Exploring the relationship between Enlightenment and Darwinist collecting and late-nineteenth-century British imperialism', *Journal of the History of Collections* XVIII/1 (2006), pp. 9–25.

9   Raj, '18th-Century pacific voyages of discovery', pp. 79–98; Margarette Lincoln, 'Mutinous behaviour on voyages to the South Seas and its impact on eighteenth-century civil society', *Eighteenth-Century Life* XXXI/1 (2007), pp. 62–80.

10  Tim Fulford, 'Romanticism, the South Seas and the Caribbean: The fruits of empire', *European Romantic Review* XI/4 (2000), pp. 408–34; Lincoln, 'Mutinous behaviour on voyages to the South Seas', pp. 62–80.

11  Janet Browne, 'A science of empire: British biogeography before Darwin', *Revue d'histoire des sciences* XLV/ 4 (1992), pp. 453–75, 464–5.

12  Barry Gough, 'The British reoccupation and colonization of the Falkland Islands, or Malvinas, 1832–1843', *Albion* XXII/2 (summer, 1990), pp. 261–87; Browne, 'A science of empire', pp. 464–5.

13  J.M.R. Cameron, 'Barrow, Sir John, first baronet (1764–1848)', *Oxford Dictionary of National Biography* (Oxford, 2004), online edition (May 2008), http://www.oxforddnb.com/view/article/1544 (accessed 6 May 2014); Browne, 'A science of empire', pp. 464–5.

14  Owen, 'Collecting artefacts, acquiring empire', pp. 9–25.

15  Jane Samson, *Imperial Benevolence: Making British Authority in the Pacific* Islands (Hawaii, 1998), pp. 11–16.

16  Ibid.

17  Owen, 'Collecting artefacts, acquiring empire', pp. 9–25; Wm. Laird Clowes, *The Royal Navy: A History from the Earliest Times to the Death of Queen Victoria*, Vol. 7 (London, 1903), p. 150; Andrew Lambert, 'Economic power, technological advantage, and imperial strength: Britain as a unique global power, 1860–1890', *International Journal of Naval History* v/2 (August, 2006).

18   Jane Samson, ' "That extensive enterprise": HMS *Herald*'s North Pacific survey, 1845–1851', *Pacific Science* LII/4 (1998), pp. 287–93, 290; Robert A. Stafford, 'Geological surveys, mineral discoveries, and British expansion, 1835–71', *Journal of Imperial and Commonwealth History*, XII/3 (1984), pp. 5–32.

19   J.F. Bosher, 'Vancouver Island in the Empire', *Journal of Imperial and Commonwealth History*, XXXIII/3 (September 2005), pp. 349–68, 354–5.

20   Steven Gray, 'Black diamonds: Coal, the Royal Navy, and British imperial coaling stations, circa 1870–1914', Ph.D. thesis, University of Warwick (March, 2014), pp. 156–68.

21   David Crane *Scott of the Antarctic: A Life of Courage, and Tragedy in the Extreme South* (London, 2005), pp. 397–9.

22   Jeremy Black, *Britain's Seaborne Empire* (London, 2004), p. 312.

# Chapter 3. *Pax Britannica*

1   Philip Ehrensaft, 'The political economy of informal empire in pre-colonial Nigeria, 1807–1884', *Canadian Journal of African Studies* VI/ 3 (1972), pp. 451–90.

2   Robert Sydney Smith, *The Lagos Consulate, 1851–1861* (Berkeley, 1979), p. 124; Kristin Mann, *Slavery and the Birth of an African City: Lagos, 1760*-1900 (Indiana, 2007), p. 100–2.

3   Ehrensaft, 'The political economy of informal empire in pre-colonial Nigeria', pp. 451–90; Marcel Van der Linden, 'Unanticipated consequences of "humanitarian intervention": The British campaign to abolish the slave trade, 1807–1900', *Theor. Soc.* XXXIX (2010), pp. 281–98.

4   Matthew S. Hopper, 'East Africa and the end of the Indian Ocean slave trade', *Journal of African Development* XIII/3 (spring, 2011), pp. 39–65.

5   Richard D. Wolff, 'British imperialism and the East African slave trade', *Science and Society* XXXVI/4 (Winter, 1972), pp. 443–62, 452–3; Raymond C. Howell, *The Royal Navy and the Slave Trade* (London, 1987); Kevin Patience, *Zanzibar, Slavery and the Royal Navy* (Bahrain, 2000); Geoffrey Ross Owens, 'Exploring the articulation of governmentality and sovereignty: The Chwaka Road and the bombardment of Zanzibar, 1895–1896', *Journal of Colonialism and Colonial History* VIII/2 (2007).

6   Owens, 'Exploring the articulation of governmentality and sovereignty'; W.H. Ingrams, *Zanzibar: Its History and Its People* (London, 1967), p. 175; Hopper, 'East Africa and the end of the Indian Ocean slave trade', pp. 39–65.

7   Richard Brooks, *The Long Arm of Empire: Naval Brigades from the Crimea to the Boxer Rebellion* (London, 1999), pp. 27–58; Clifford Pereira, 'Black liberators: The royal of Africans and Arab sailors in the Royal Navy within the Indian Ocean 1841–1941', research paper presented at the UNESCO symposium in Rabat, Morocco (May, 2007).

8     Pereira, 'Black Liberators'; George E. Brooks Jr, *The Kru Mariner in the Nineteenth Century: An Historical Compendium*, Liberian Studies Monograph Series No. 1 (Newark, 1972); Ronald W. Davis, *Ethnohistorical Studies on the Kru Coast*, Liberian Studies Monograph Series No. 5 (Newark, 1976); Diane Frost, *Work and Community among West African Migrant Workers since the Nineteenth Century* (Liverpool, 1999).

9     Pereira, 'Black Liberators'; Janet J. Ewald, 'Crossers of the sea: Slaves, freedmen, and other migrants in the northwestern Indian Ocean, *c*.1750–1914', *American Historical Review*, cv/1 (February, 2000), pp. 69–91; Jayasuriya Shihan de Silva and Richard Pankhurst, *The African Diaspora in the Indian Ocean* (Trenton, 2003).

10    Pereira, 'Black Liberators'; Donald Simpson, *Dark Companions: The African Contribution to the European Exploration of East Africa* (London, 1975); Howell, *The Royal Navy and the Slave Trade*.

11    F.M. Withers, 'Nyasaland in 1895–96', *Nyasaland Journal* II / 1 (January, 1949), pp. 16–34.

12    David Killingray, 'The idea of a British imperial African Army', *Journal of African History* XX/3 (1979), pp. 421–36, 422; Pereira, 'Black Liberators'; Marika Sherwood, 'More on blacks in the Royal Navy', *Black and Asian Studies Association Newsletter* XXVI (January, 2000), pp. 8–10.

13    Barry M. Gough, 'Send a gunboat! Checking slavery and controlling liquor traffic among coast Indians of British Columbia in the 1860s', *The Pacific Northwest Quarterly* LXIX/4 (October, 1978), pp. 159–68.

14    L.F.S. Upton, 'Contact and conflict in the Atlantic and Pacific coasts of Canada', *Acadiensis* IX/2 (Spring, 1980), pp. 3–13, 9–12.

15    Daniel R. Headrick, 'The tools of imperialism: Technology and the expansion of European colonial empires in the nineteenth century', *The Journal of Modern History* LI/2 (June 1979), pp. 231–63; Cindy McCreery, 'Telling the story: HMS Galatea's voyage to South Africa, 1867', *South African Historical Journal* LXI/4 (2009), pp. 817–37.

16    Commander J. Elliot Bingham, RN, *Narrative of the Expedition to China, from the Commencement of the War to its Termination in 1842; With Sketches of the Manners and Customs of that Singular and Hitherto Almost Unknown Country*, Vol. 1 (London, 1843), p. 112–13.

17    Headrick, 'The tools of imperialism', p. 242.

18    Commander William Hutcheon Hall, RN, and William Dallas Bernard, *Narrative of the Voyages and Services of the Nemesis from 1840 to 1843; And of the Combines Naval and Military Operations in China*, 2nd ed. (London, 1845), p. 126.

19    Headrick, 'The tools of imperialism', pp. 243–4.

20    Francis Rosebro Flourney, *British Policy towards Morocco in the Age of Palmerston* (London, 1935).

21 John Gallagher and Ronald Robinson, 'The Imperialism of Free Trade', *Economic History Review* vi/1 (1953), p. 8.

22 John Lynch, 'British policy and Spanish America, 1783–1808', *Journal of Latin American Studies* I/1 (May, 1969), pp. 1–30.

23 David Mclean, 'Trade, politics and the Navy in Latin America: The British in the Parana, 1845–46', *Journal of Imperial and Commonwealth History* XXXV/3 (August 2007), pp. 351–70, p. 360.

24 Barry Gough, 'The British Reoccupation and Colonization of the Falkland Islands, or Malvinas', 1832–1843', *Albion* XXII/2 (Summer, 1990) p. 263.

25 Jane Samson, *Imperial Benevolence: Making British Authority in the Pacific* Islands (Hawaii, 1998), pp. 11–16.

26 John Bach 'The Royal Navy in the Pacific Islands', *The Journal of Pacific History* III (1968), pp. 3–20; Holland to Carnarvon, 20 August 1874, Carnarvon Papers Mss 607968.

27 Brooks, *The Long Arm of Empire*, pp. 111–18.

28 Satyindra Singh, *Blueprint to Bluewater: The Indian Navy 1951–65* (New Delhi, 1992), pp. 257–61.

29 Oliver B. Pollak, 'The origins of the second Anglo-Burmese War (1852–53)', *Modern Asian Studies* XII/3 (1978), pp. 483–502; Palmerston, 25 June 1850, Hansard CXII [3d Ser.], 380–444.

30 Pollak, 'The origins of the second Anglo-Burmese War', pp. 483–502.

31 Brooks, *The Long Arm of Empire*, pp. 27–58.

32 Robert Holland, *Blue-Water Empire: The British in the Mediterranean since 1800* (London, 2012), p. 110–14.

33 T.D. Potgieter, 'Maritime defence and the South African Navy to the cancellation of the Simon's Town Agreement', *Scientia Militaria: South African Journal of Military Studies*, XXX/2 (2000), pp. 165–6; Brooks, *The Long Arm of Empire*, pp. 214–36.

34 Brooks, *The Long Arm of Empire*, pp. 216–35.

35 Mary Conley, *From Jack Tar to Union Jack: Representing Naval Manhood in the British Empire, 1870–1918* (Manchester, 2009), pp. 142–9.

36 Erik J. Dahl, 'Naval innovation: From coal to oil', *Joint Force Quarterly*, XXVII (January, 2001), pp. 50–6.

37 G.G. Jones, 'The state and economic development in India 1890–1947: the case of oil', *Modern Asian Studies*, XIII/3 (1979), pp. 353–75.

38 Quoted in Dahl, 'Naval innovation: From coal to oil', pp. 50–6.

39 Marian Jack, 'The purchase of the British government's shares in the British Petroleum Company 1912–1914', *Past and Present* XXXIX (April, 1968), pp. 139–68.

## Chapter 4. Imperial and Colonial Culture

1   George Keate, 'Verses to Captain Bligh', in James Stanier Clarke, John Jones, and Stephen Jones (eds), *The Naval Chronicle*, Vol. 4 (London, 1801), pp. 495–6.

2   Lynda Pratt, 'Naval contemplation', *Journal for Maritime Research* XXII/1 (2000), pp. 84–105.

3   Tim Fulford, 'Romanticizing the Empire: The Naval Heroes of Southey, Coleridge, Austen and Marryat', *Modern Language Quarterly*, Vol. 60, No. 2 (1999), pp. 161–196.

4   Ibid.

5   Mark Connelly, 'Battleships and British society, 1920–1960', *International Journal of Naval History* III/2–3 (August/December, 2004), p. 8; Fulford, 'Romanticizing the empire', pp. 161–96.

6   Francois Cellier and Cunningham Bridgeman, *Gilbert and Sullivan and Their Operas* (London, 1914), p. 60; Robert Holland, *Blue-Water Empire : The British in the Mediterranean since 1800* (London, 2012), pp 110–14.

7   Derek Anthony Swain, 'A history of sport in British Columbia to 1885: A chronicle of significant developments and events', Master's thesis, University of British Columbia (April, 1977); J.F. Bosher, 'Vancouver Island in the Empire', *Journal of Imperial and Commonwealth History*, XXXIII/3 (September 2005), pp. 349–68.

8   Tadasu Tsuruta, 'Popular music, sports, and politics: A development of urban cultural movements in Dar es Salaam, 1930s–1960s', *African Study Monographs* XXIV/3 (2003), pp. 195–222; Margaret Strobel, 'Women's wedding celebrations in Mombasa, Kenya', *African Studies Review* XVIII/3, 'Women in Africa' (December, 1975), pp. 35–45.

9   Laura Fair, ' "It's just no fun anymore": Women's experiences of Taarab before and after the 1964 Zanzibar Revolution', *The International Journal of African Historical Studies* XXXV/1 (2002), pp. 61–81.

10  Brigitte Reinwald, ' "Tonight at the empire": Cinema and urbanity in Zanzibar, 1920s to 1960s', *Verdier: Afrique & histoire* v/1 (2006), pp. 81–109.

11  The National Archives, ADM1/23215, 'A short story of the Trinidad Royal Naval Reserve', pp. 3–4.

12  Errol G. Hill, 'Calypso and War', *Black American Literature Forum* XXIII/1 (Spring, 1989), p. 68.

13  Ibid., pp. 61–2; Gordon Rohlehr, 'I Lawa: The construction of masculinity in Trinidad and Tobago Calypso', in Rhoda E. Reddock, *Interrogating Caribbean Masculinities: Theoretical and Empirical Analyses* (University Press of the West Indies, 2004), pp. 326–8; Daniel Owen Spence, *Colonial naval culture and British imperialism, 1922-67* (Manchester, 2015), chapter 2.

14  Alex Law, 'Of navies and navels: Britain as a mental islands', *Geografiska Annaler: Series B, Human Geography* LXXXVII/4 (December, 2005), pp. 267–77; Mary

Conley, *From Jack Tar to Union Jack: Representing Naval Manhood in the British Empire, 1870–1918* (Manchester, 2009), pp. 142–9.

15 Swain, 'A history of sport in British Columbia to 1885'.

16 Ibid.

17 Ibid.

18 Gareth Stockey, 'Sport in Gibraltar – problematizing a supposed "problem"', 1713–1954', *Sport in History* XXXII/1 (2012), pp. 1–25.

19 Leigh Jones and Nicky Jones, 'Emergence, cessation and resurgence during the evolution of Rugby Union in Hong Kong', *The International Journal of the History of Sport* XXIX/9 (2012), pp. 1344–62; *South China Morning Post*, 30 September 1933, p. 16; Spence, *Colonial naval culture*, chapter 7.

20 The National Archives, ADM 1/9749, 'From Commanding Officer, H.M.S. "DRAGON" at Minatitlan, to the Commander-in-Chief, America and West Indies', 2 February 1937; ADM 1/9749, 'Cayman Islands regatta - proposed presentation of a cup', 4 May 1937; Spence, *Colonial Naval Culture*, chapter 3; Michael Craton, *Founded Upon the Seas: A History of the Cayman Islands and Their People* (Kingston, 2003), p. 291.

21 Cayman Islands National Archive, interview with Harry McCoy by Heather McLoughlin, 3 July 1991, tape 2A, p. 10; The National Archives, CO 537/1891, Trinidad RNVR: Training, Item 9; Spence, *Colonial naval culture*, chapters 2–3.

22 Zanzibar National Archives, BF 15/3, Report: Royal East African Navy, 1955–56.

23 Zanzibar National Archives, AB 49/127, Royal East African Navy Report, 1952; BF15/4, Report: Royal East African Navy, 1957; BF15/6, Report: Royal East African Navy, 1959; AB49/130, Commander E.A. Nicholson, to British Resident, Zanzibar, 14 January 1955; Spence, *Colonial naval culture*, chapter 5.

24 Zanzibar National Archives, AB 49/127, Royal East African Navy Report, 1953; BF15/3, Report: Royal East African Navy, 1955–56; BF15/7, Report: Royal East African Navy, 1960; BF15/8, Report: Royal East African Navy, 1961; Kenya National Archives, DC/KSM/1/22/34, Information Officer, Nyanza, to District Commissioners North, Central, South Nyanza, and Kricho, 19 March 1955; Spence, *Colonial naval culture*, chapter 5.

25 Interview with Reverend Neivelle Tan conducted by the author on 14 July 2009 in Singapore; N.G. Aplin and Quek Jin Jong, 'Celestials in touch: Sport and the Chinese in colonial Singapore', *The International Journal of the History of Sport* XIX/2–3 (2002), pp. 67–98.

26 Interview with Lieutenant Commander Karu Selvaratnam conducted by the author on 30 June 2009 in Kuala Lumpur; Royal Malaysian Navy (ed.), *Serving the Nation* (Kuala Lumpur, 2004), p. 40.

## Chapter 5. Colonial Navalism

1   Sir Charles Lucas (ed.) *The Empire At War*, Vol. I (Oxford, 1921), p. 77; Bob Nicholls, 'Colonial naval forces before Federation', in David Stevens and John Reeve (eds), *Southern Trident: Strategy, History and the Rise of Australian Naval Power* (Crows Nest, 2001), pp. 125–39.

2   'Report of the Royal Commissioners Appointed to Enquire into the Defence of British Possessions and Commerce Abroad (1882)' in Naval Archives Branch, *An Outline of Australian Naval History* (Canberra, 1962), p. 13; W.T. Stead, 'What is the truth about the navy', *Pall Mall Gazette*, 15 September 1884, The *W.T. Stead Resource Site*, http://www.attackingthedevil.co.uk/pmg/navy.php (accessed 21 December 2014).

3   Richard Jebb, *The Imperial Conference*, Vol. I (London, 1911), pp. 41–2; H.C. Ferraby, *The Imperial British Navy: How the Colonies Began to Think Imperially Upon the Future of the Navy* (London, 1918); Daniel Owen Spence, 'Australian Naval Defence and the 1887 Colonial Conference: Context, Policy, and Reaction', *International Journal of Naval History* VI/1 (April, 2007).

4   Richard H. Gimblett, 'Reassessing the Dreadnought Crisis of 1909 and the origins of the Royal Canadian Navy', *The Northern Mariner*, IV/1 (January, 1994), pp. 35–53, 36.

5   Roger Sarty, 'Canadian maritime defence, 1892–1914', *Canadian Historical Review* LXXI/4 (1990), pp. 462–90.

6   Phillip Buckner, 'The Royal tour of 1901 and the construction of an imperial identity in South Africa', *South African Historical Journal* XLI/1 (1999), pp. 324–48.

7   Ibid.

8   Cindy McCreery, 'Telling the story: HMS Galatea's voyage to South Africa, 1867', *South African Historical Journal* LXI/4 (2009), pp. 817–7.

9   Ibid.

10  Jan Rüger, *The Great Naval Game: Britain and Germany in the Age of Empire* (Cambridge, 2007), pp. 175–82; Nicholas Tracy (ed.), *The Collective Naval Defence of the Empire 1900–1944* (Aldershot, 1997), p. xii; Ferraby, *The Imperial British Navy*.

11  Rüger, *The Great Naval Game*, pp. 175–82.

12  Richard Brooks, *The Long Arm of Empire: Naval Brigades from the Crimea to the Boxer Rebellion* (London, 1999), pp. 236–48.

13  Ibid.

14  Richard H. Wilde, 'The Boxer affair and Australian responsibility for imperial defense', *Pacific Historical Journal* XXVI/1 (February, 1957), pp. 51–65.

15  Lord Selbourne, Memorandum 'The Colonies and the Navy', 20 May 1902, in Tracy (ed.), *The Collective Naval Defence of the Empire*, pp. 8, 12.

16  Phillips Payson O'Brien, 'The titan refreshed: Imperial overstretch and the British Navy before the First World War', *Past and Present*, 172 (August, 2001), pp. 146–69; Harold Mackinder, *Britain and the British Seas* (London, 1902), p. 358; Brian W. Blouet, 'The Imperial Vision of Harold Mackinder', *The Geographical Journal* CLXX/4 (December, 2004), pp. 322–9; Donald C. Gordon, 'The Admiralty and dominion navies, 1902–1914', *The Journal of Modern History* XXXIII/4 (December, 1961), pp. 407–22; Ferraby, *The Imperial British Navy*.

17  *The Lone Hand*, 1 August 1908, quoted in Sea Power Centre, 'The Great White Fleet's 1908 Visit to Australia', *Semaphore*, 8 (July, 2008); Gordon, 'The Admiralty and dominion navies', p. 411; Alfred Deakin quoted in Arthur W. Jose, *The Royal Australian Navy* (Queensland, 1987), pp. 57–8.

18  Archibald Hurd, 'The racial war in the Pacific: An imperial peril', *Fortnightly Review* XCIII/558 (June, 1913), p. 1032

19  John Augustus Barron, The Empire Club of Canada Addresses, 11 March 1909, pp. 158–67, http://speeches.empireclub.org/62348/data?n=11 (accessed 21 December 2014).

20  Sarty, 'Canadian maritime defence', p. 477; Gimblett, 'Reassessing the Dreadnought Crisis of 1909', p. 46; Ferraby, *The Imperial British Navy*.

21  Sir Joseph Ward, letter to Reginald McKenna, 11 August 1909, in Tracy (ed.), *The Collective Naval Defence of the Empire*, p. 109.

22  Connelly, 'Battleships and British society'.

23  Ferraby, *The Imperial British Navy*, pp. 90–3; John C. Mitcham, 'Navalism and Greater Britain, 1897–1914', in Duncan Redford (ed.), *Maritime History and Identity: The Sea and Culture in the Modern World* (London, 2013), pp. 271–93.

24  Gordon, 'The Admiralty and dominion navies', pp. 407–22.

25  *The Times*, 14 June 1909, p. 9; Simon J. Potter, *News and the British World: The Emergence of an Imperial Press System* (Oxford, 2003), pp. 136–7; Rüger, *The Great Naval Game*, pp. 175–82.

26  Gordon, 'The Admiralty and dominion navies', pp. 412–14.

27  Ibid.

28  Bernard Ransom, 'A nursery of fighting seamen? The Newfoundland Royal Naval Reserve, 1901–1920', in Michael L. Hadley, Rob Huebert and Fred W. Crickard (eds), *Nation's Navy: In Quest of Canadian Naval Identity*, pp. 239–52; Ferraby, *The Imperial British Navy*.

29  Mitcham, 'Navalism and Greater Britain', p. 278; Royal New Zealand Navy Museum, Torpedo Bay, visited on 6 December 2014.

30  Ferraby, *The Imperial British Navy*.

31  Royal New Zealand Navy Museum, Torpedo Bay, visited on 6 December 2014; Mitcham, 'Navalism and Greater Britain', p. 278.

32  O'Brien, 'The titan refreshed', p. 155.

33   Rüger, *The Great Naval Game*, pp. 175–82. Ferraby, *The Imperial British Navy*;
     Jose, *The Royal Australian Navy*.
34   T.D. Potgieter, 'Maritime defence and the South African Navy to the
     cancellation of the Simon's Town Agreement', *Scientia Militaria: South African
     Journal of Military Studies*, XXX/2 (2000), pp. 166–7; Ferraby, *The Imperial
     British Navy*.

## Chapter 6. The Empire Rallies

1    T.D. Potgieter, 'Maritime defence and the South African navy to the cancellation
     of the Simon's Town Agreement', *Scientia Militaria: South African Journal of
     Military Studies*, XXX/2 (2000)', pp. 167–8.
2    'Papers Relating to the Foreign Relations of the United States: 1922',
     Vol. 1, pp. 252–3, http://www.ibiblio.org/pha/pre-war/1922/nav_lim.html
     (accessed on 21 December 2014).
3    The National Archives, ADM 1/8715/189, Admiralty to the Colonial Office, 22
     July 1927, p. 2; David W. McIntyre, 'New Zealand and the Singapore base
     between the wars', *Journal of Southeast Asian Studies* II/1 (March, 1971), pp. 2–21.
4    The National Archives, ADM 116/3125, 'Remarks on Paragraphs in Detail,
     Part II', p. 5; CO 323/902, Item 121, Committee of Imperial Defence to
     Admiralty, 21 March 1923; Paul Haggie, *Britannia At Bay: The Defence of the
     British Empire against Japan, 1931–1941* (Oxford, 1981), p. 7.
5    The National Archives, CAB 21/187, Empire Naval Policy and Cooperation,
     February 1921, p. 44.
6    The National Archives, ADM 116/2219, to Secretary, Treasury, 4 August 1923;
     Ralph Harrington, ' "The mighty hood": Navy, empire, war at sea and the
     British national imagination, 1920–60", *Journal of Contemporary History*
     XXXVIII/2 (April, 2003), pp. 171–5; V.C. Scott O'Connor, *The Empire
     Cruise* (London, 1925).
7    Harrington, 'The mighty hood'; Mark Connelly, 'Battleships and British society,
     1920–1960', *International Journal of Naval History*, III/2–3 (August/December,
     2004)'.
8    Ibid.; Tom Rice, 'Britain's Birthright, 1924', BFI 133391, Colonial Film Archive
     (April, 2008), http://www.colonialfilm.org.uk/node/6219 (accessed 31
     October 2012); Ashley Jackson, *The British Empire and the Second World War*
     (London, 2006), p. 57; Jon Wise, *The Role of the Royal Navy in South America,
     1920–1970* (London, 2014), pp. 20–2.
9    C.J. Duder, ' "Men of the officer class": The participants in the 1919 soldier
     settlement scheme in Kenya', *African Affairs* XCII/366 (January, 1993), pp. 69–
     87; Carl G. Rosberg Jr and John Nottingham, *The Myth of 'Mau Mau':
     Nationalism in Kenya* (London, 1966), pp. 65–9.

10   Royal New Zealand Navy Museum, Torpedo Bay, Devonport, visited on 6 December 2014.

11   Roger Dingman, *Power in the Pacific: The Origins of Naval Arms Limitation, 1914– 1922* (Chicago, 1976), p. 109; Andrew Field, *Royal Navy Strategy in the Far East, 1919–1939: Preparing for War against Japan* (London, 2004), pp. 14, 43; C. Robinson and H.M. Ross (eds), *Brassey's Naval and Shipping Annual* (London, 1931); Haggie, *Britannia At Bay*, p. 9; Daniel Owen Spence, *Colonial naval culture and British imperialism, 1922-67* (Manchester, 2015), chapter 1.

12   Original underlining. The National Archives, ADM 1/8715/189, 'Extract from 278th Minutes of the Overseas Defence Committee, 23 May 1927', pp. 1–2; ADM 1/8715/189, 'Contributions by Colonies towards cost of Naval Defence', 12 August 1927, p. 2.

13   Stephen Lyon Endicott, *Diplomacy and Enterprise: British China Policy, 1933–1937* (Manchester, 1975), p. 184; Ian Nish, 'Japan in Britain's view of the international system, 1919–37', in Ian Nish (ed.), *Anglo-Japanese Alienation, 1919–1952: Papers of the Anglo-Japanese Conference on the History of the Second World War* (Cambridge, 1982), pp. 43–4, 53; Arthur J. Marder, *Old Friends, New Enemies: The Royal Navy and the Imperial Japanese Navy, Strategic Illusions, 1936– 1941* (Oxford, 1981), p. 35; Spence, *Colonial naval culture*, chapter 1.

14   Field, *Royal Navy Strategy in the Far East*, pp. 43, 99; Potgieter, 'Maritime defence and the South African navy', pp. 167–8; Marder, *Old Friends, New Enemies*, p. 11.

15   Nicholas Tracy (ed.), *The Collective Naval Defence of the Empire 1900–1944* (Aldershot, 1997), pp. xl–li, 26.

16   Item 340: Memorandum 'The co-operation of the colonial empire in imperial defence', signed by V. Sykes, Secretary, Overseas Defence Committee, 8 June 1938, in Tracy, *The Collective Naval Defence of the Empire*, pp. 587–93; Spence, *Colonial naval culture*, chapter 1.

17   Connelly, 'Battleships and British society'.

18   Rice, 'Britain's Birthright'.

19   Jackson, *The British Empire and the Second World War*, pp. 55–85; Joseph Schull, *Far Distant Ships: An Official Account of Canadian Naval Operations in World War II* (Toronto, 1987), p. 430.

20   Howard Johnson, 'Oil, imperial policy and the Trinidad disturbances, 1937', *Journal of Imperial and Commonwealth History* IV/ 1 (1975), pp. 29–54; Vernon C. Mulchansingh, 'The oil industry in the economy of Trinidad', *Caribbean Studies* XI/1 (April, 1971), pp. 73–100, 73; Fitzroy A. Baptiste, 'The European possessions in the Caribbean in World War II: dimensions of great power co-operation and conflict', Ph.D. thesis, University of the West Indies, Trinidad (1981), p. 21; The National Archives, ADM 1/10969, Director of L.D., 18 September 1939; CO 537/1891, 'Trinidad Royal Naval Volunteer Reserve', 1946, p. 2; ADM 1/23215, Senior Naval Officer Trinidad's communication, 3 August 1940; ADM 1/23215, 'Appreciation of Naval Organisation In The

West Indies', 3 June 1945, Appendix "A"; ADM 1/23215, C-in-C America and West Indies to Admiralty, 2 November 1951; The Cayman Islands National Archives, *The Northwester*, March 1974, pp. 6–8; Gaylord Kelshall, *The U-Boat War in the Caribbean* (Annapolis, 1994), pp. foreword–1; Oliver Marshall (ed.), *The Caribbean at War, British West Indians in World War II* (London, 1992), p. 25; Spence, *Colonial Naval Culture*, chapter 2.

21 Jackson, *The British Empire and the Second World War*, pp. 55–85.

22 The National Archives, CO 968/145/2, 'Future of Colonial Naval Forces', 1945; Jackson, *The British Empire and the Second World War*, pp. 227–8, 122–35.

23 'Able Seaman Just Nuisance, RN', The Simon's Town Museum, 2001, http://www.simonstown.com/tourism/nuisance/nuisance.ht (accessed on 20 December 2014); Potgieter, 'Maritime defence and the South African Navy', pp. 159–82; Connelly, 'Battleships and British Society', pp. 289–30; Jackson, *The British Empire and the Second World War*, pp. 241–52, 343.

24 The National Archives, ADM 1/18245, 'Kenya Royal Naval Volunteer Reserve' (Enclosure to Captain-in-Charge, Kilindini), 23 March 1946; CO 968/80/8, 'East African Naval Defence Forces', 1943; CO 968/145/2, 'Future of colonial naval forces', 1945; ADM 1/16071, Tanganyika Territory (Sir W. Jackson) to Secretary of State, Colonies, 7 June 1944; Zanzibar National Archives, AB 66/1, Commander-in-Chief, East Indies, to British Resident, Zanzibar, 5 July 1939; AB 66/1, Zanzibar to Lords of Admiralty, 16 July 1891; AB 66/4, 'Zanzibar Naval Volunteer Force', C.G. Somers, 26 September 1938; AB 66/4, 'Raising of Zanzibar Naval Volunteer Force'; Spence, *Colonial naval culture*, chapter 4.

25 The National Archives, ADM1/18245, Commander-in-Chief East Indies, 'Kenya Royal Naval Volunteer Reserve', 15 December 1945; Spence, *Colonial Naval Culture*, chapter 4; Jeremy Black, *The British Seaborne Empire* (London, 2004), p. 312; Jackson, *The British Empire and the Second World War*, pp. 147–215, 357–69.

26 The National Archives, CO 968/24/2, Governor to Secretary-of-State for the Colonies, 18 March 1942; Ashley Jackson, *War and Empire in Mauritius and the Indian Ocean* (Basingstoke, 2001), pp. 10, 161, 173–4, 177.

27 The National Archives, CO 968/145/2, 'Future of colonial naval forces', 1945; CO 129/584/8, Hong Kong Royal Naval Volunteer Reserve Report, 1940; Jackson, *The British Empire and the Second World War*, pp. 452–6.

28 Public Records Office, Hong Kong, HKRS 6-1-1706, J.C. McDouall Papers, p. 38; HKRS 6-1-1706, J.C. McDouall to A. Sommerfelt, 5 April 1961, p. 1; HKRS 7-1-1716, Captain, Extended Defence Officer to Commodore, Hong Kong, 3 May 1941; Sir Selwyn Selwyn-Clarke, *Footprints: The Memoirs of Sir Selwyn Selwyn-Clarke* (Hong Kong, 1975), p. 58.

29 Public Records Office, Hong Kong, HKRS 6-1-1706, 'With the MTBs – December 1941', letter of a Sub-Lieutenant, Hong Kong Royal Naval Volunteer Reserve, to his brother; Tim Luard, *Escape from Hong Kong: Admiral Chan Chak's Christmas Day Dash, 1941* (Hong Kong, 2012), Kindle edition location 1038–53;

Phillip Snow, *The Fall of Hong Kong: Britain, China and the Japanese Occupation* (London, 2003), p. 73; R.B. Goodwin, *Hongkong Escape* (London, 1953), pp. 220–1; Spence, *Colonial naval culture*, chapter 7.

30 Royal Malaysian Navy Museum, Malacca, visited on 2 July 2009; Dato Yusof Nordin (ed.), *Honour and Sacrifice* (Kuala Lumpur, 1994), p. 21; *The Straits Times*, 7 December 1934, p. 13, 8 December 1935, p. 2, 28 December 1939, p. 10; Spence, *Colonial naval culture*, chapter 5.

31 Winston Churchill, *The Grand Alliance* (London, 1950), p. 487; Mahmud Ghazemy and Mohd. Haidar Abu Bakar (eds.), *Royal Malaysian Navy: 55 Years* (Kuala Lumpur, 1990), p. 77; *The Straits Times*, 7 July 1946, p. 4.

32 Interview with Ahmad, Dahim bin Noordin, by Daniel Chew, 5 March 1992, 001318, reel 3, National Archives of Singapore.

33 The National Archives, CO 968/145/2, 'Future of colonial naval forces', 1945; ADM 1/19551, Burma Navy; ADM 1/12995, Sabben-Clare to M.J. Fox, 27 July 1943; Jackson, *The British Empire and the Second World War*, pp. 292–4, 308–30, 398–9, 459–60.

34 The National Archives, CO 968/145/2, 'Future of colonial naval forces', 1945; Royal New Zealand Navy Museum, Torpedo Bay, visited on 6 December 2014; Jackson, *The British Empire and the Second World War*, pp. 381–503.

35 The National Archives, FO 371/46251, 'Arrangements for the Administration of Hong Kong in the Event of its Liberation by Regular Chinese Forces', July 1945, quoted in Steve Tsang, *A Modern History of Hong Kong* (London, 2007), p. 133.

36 The National Archives, CO 968/80/7, 'Disturbances at Trinidad Royal Naval Volunteer Reserve Headquarters', 9 October 1943, p. 1; ADM 178/301, Appendix to Board of Inquiry Proceedings, 1943; Norman Rudolph McLaughlin, *The Forgotten Men of the Navy* (Miami, 2002), pp. 8–16, 119; Spence, *Colonial naval culture*, chapters 2–3; India Office Records, L/MIL/17/9/379, 'Report of the Royal Indian Navy Commission of Enquiry', 1946, p. 460; Daniel Owen Spence, 'Imperial transition, indianisation and race: developing national navies in the subcontinent, 1947–1964', *South Asia: Journal of South Asian Studies* XXXVII/2 (June, 2014), pp. 323–38.

## Chapter 7. End of Empire?

1 Jan Rüger, *The Great Naval Game: Britain and Germany in the Age of Empire* (Cambridge, 2007), pp. 270–1; D.A. Low and J. Lonsdale, 'Introduction', in D.A. Low and Alison Smith (eds), *The Oxford History of East Africa* (Oxford, 1976), pp. 1–64.

2 Jeremy Black, *The British Seaborne Empire* (London, 2004), p. 337; Cynthia H. Enloe, 'Ethnicity in the evolution of Asia's armed bureaucracies', in C. Ellinwood DeWitt and Cynthia H. Enloe (eds), *Ethnicity and the Military in Asia* (London,

1981), pp. 1–17, 9; Peter Fosten, 'A Far East odyssey, part two: Aiding the civil power, Malaya 1950–53', unpublished memoir acquired from the author in Newton Abbot on 14 August 2009; Interview with Captain Chitharanjan Kuttan, by Jason Lim, September 2002, 002697, reel 4, National Archives of Singapore.

3    Daniel Owen Spence, 'Imperial Transition, Indianisation and Race: Developing National Navies in the Subcontinent, 1947–1964', *South Asia: Journal of South Asian Studies* XXXVII/2 (June, 2014)'; Interview with Admiral Tan Sri Thanabalasingam Karalasingam conducted by the author on 25 June 2009 in Kuala Lumpur.

4    Ashley Jackson, 'The Royal Navy and the Indian ocean region since 1945', *The RUSI Journal* CLI/6 (2006), pp. 78–82.

5    Fraser MacDonald 'The last outpost of Empire: Rockall and the Cold War', *Journal of Historical Geography* XXXII (2006), pp. 627–47.

6    Black, *The British Seaborne Empire*, p. 335; Jackson, 'The Royal Navy and the Indian ocean region', pp. 78–82.

7    Royal New Zealand Navy Museum, Torpedo Bay, Devonport, visited on 6 December 2014.

8    Ian Speller, 'The seaborne/airborne concept: littoral manoeuvre in the 1960s?', *Journal of Strategic Studies* XXIX/1 (2006), pp. 5382; Black, *Britain's Seaborne Empire*, p. 327.

9    Jonathan Rayner ' "Entrusted with the ruling of the waves": images of the post-war royal navy in the NMM film archive', *Journal of Maritime Research* x/1 (2008), pp. 50–66; W. Taylor Fain, *American Ascendance and the British Retreat in the Persian Gulf Region* (Palgrave, 2008), p. 3.

10   Rayner, 'Entrusted with the ruling of the waves'; Duncan Redford, 'The "hallmark of a first-class navy" the nuclear-powered submarine in the Royal Navy 1960–77', *Contemporary British History* XXIII/2 (May, 2009), pp. 181–97; See also Duncan Redford, *The Submarine: A Cultural History from the Great War to Nuclear Combat* (London, 2010).

11   John Subritzky 'Britain, *Konfrontasi*, and the end of empire in Southeast Asia, 1961–65', *Journal of Imperial and Commonwealth History* XXVIII/3 (2000), pp. 209–27, 211.

12   Daniel Owen Spence, *Colonial naval culture and British imperialism, 1922-67,* (Manchester, 2015), chapter 5; Ian Speller, 'An African Cuba? Britain and the Zanzibar Revolution, 1964', *Journal of Imperial and Commonwealth History*, Vol. 35, No. 2 (June 2007), pp. 283-301; The National Archives, CO 822/562, Rankine, to E.B. David, 8 July 1953; CO 822/759, Goodenough, to Rankine, 7 April 1954.

13   Spence, 'Imperial Transition, Indianisation and Race'; Christopher MacRae and Tony Laurence, 'The 1964 Tanganyika Rifles mutiny and the British armed intervention that ended it', *The RUSI Journal* CLII/2 (2007), pp. 96–101; The National Archives, CAB 148/4, Kenya: Defence and Financial Discussions,

12 March 1964; Martin Bailey, 'Tanzania and China', *African Affairs* LXXIV/294 (January, 1975), pp. 39–50.

14  Richard Mobley, 'The Beira Patrol: Britain's Broken Blockade against Rhodesia', *Naval War College Review* LV/1 (winter, 2000), pp. 63–84.

15  T.D. Potgieter, 'Maritime defence and the South African Navy to the cancellation of the Simon's Town Agreement', *Scientia Militaria: South African Journal of Military Studies*, XXX/2 (2000)', pp. 175–82.

16  Jackson, *War and Empire in Mauritius and the Indian Ocean* (Basingstoke 2001), pp. 181–82; Black, *The British Seaborne Empire*, p. 333.

17  Andrew Cohen, 'Lonrho and oil sanctions against Rhodesia in the 1960s', *Journal of Southern African Studies* XXXVII/4 (2011), pp. 715–30; Mobley, 'The Beira Patrol'.

18  Black, *The British Seaborne Empire*, p. 347.

19  Arthur Hermann, *To Rule the Waves: How the British Navy Shaped the Modern World* (Chatham, 2005), p. 560; *Newsweek*, 19 April 1982.

20  Quoted in James R. Andrews, 'The imperial style: rhetorical depiction and queen Victoria's diamond jubilee', *Western Journal of Communication* LXIV/1 (2000), pp. 53–77, 54.

21  Charles Dilke, 1872, quoted in Peter J. Cain, 'Empire and the languages of character and virtue in later Victorian and Edwardian Britain', *Modern Intellectual History* IV/2 (August, 2007), pp. 249–73; Black, *The British Seaborne Empire*, p. 347.

22  Black, *The British Seaborne Empire*, p. 317.

23  Spence, *Colonial naval culture*, chapter 3; Craton *Founded Upon the Seas: A History of the Cayman Islands and Their People* (Kingston, 2003).

24  'Special Committee on Decolonization "no longer relevant" to overseas territories of United Kingdom, fourth committee told', United Nations, 11 October 2012, http://www.un.org/press/en/2012/gaspd507.doc.htm (accessed 5 December 2014).

25  Robert Holland, *Blue-Water Empire: The British in the Mediterranean since 1800* (London, 2012), pp. 348–9.

26  *The Telegraph*, 10 October 2010, http://www.telegraph.co.uk/news/world-news/southamerica/falklandislands/8053835/Royal-Navy-are-pirates-says-Argentinas-president.html; 17 June 2011, www.telegraph.co.uk/news/world-news/southamerica/falklandislands/8581447/Britain-a-crude-colonial-power-in-decline-says-Argentinas-president-Cristina-Kirchner.html; and 18 January 2012, http://www.telegraph.co.uk/news/worldnews/southamerica/falklan-dislands/9022752/David-Cameron-accuses-Argentina-of-colonialism-over-Falklands-Islands.html (accessed on 20 December 2014).

# Bibliography

## Archives

Cayman Islands National Archives
India Office Records, British Library
Kenya National Archives
Museum of Australian Democracy
National Archives of Singapore
National Museum of the Royal Navy
Public Records Office, Hong Kong
Royal Malaysian Navy Museum
Royal New Zealand Navy Museum
The National Archives, United Kingdom
Zanzibar National Archives

## Articles

Andrews, James R., 'The imperial style: Rhetorical depiction and Queen Victoria's diamond jubilee', *Western Journal of Communication* LXIV/1 (2000), pp. 53–77.

Aplin, N.G. and Quek Jin Jong, 'Celestials in touch: Sport and the Chinese in colonial Singapore', *The International Journal of the History of Sport* XIX/2–3 (2002), pp. 67–98.

Bach, John, 'The Royal Navy in the Pacific Islands', *The Journal of Pacific History*, III (1968), pp. 3–20.

Bailey, Martin, 'Tanzania and China', *African Affairs* LXXIV/294 (January, 1975), pp. 39–50.

Barnes, Geraldine and Adrian Mitchell, 'Measuring the marvelous: science and the exotic in William Dampier', *Eighteenth-Century Life* XXVI/3 (fall, 2002), pp. 45–57.

Baugh, Daniel A., 'Great Britain's "blue-water" policy, 1689–1815', *The International History Review* x/1 (February, 1988), pp. 33–58.

——'Maritime strength and Atlantic commerce: The uses of "a grand marine empire", in Lawrence Stone (ed.), *An Imperial State at War: Britain From 1689 to 1815* (London, 1994), pp. 185–203.

Blouet, Brian W., 'The imperial vision of Harold Mackinder', *The Geographical Journal* CLXX/4 (December, 2004), pp. 322–9.

Borgie, Francisco J., 'Richard Hakluyt, promoter of the New World: the navigational origins of the English nation', *Sederi* XIII (2003), pp. 1–9.

Boruchoff, David A., 'Piety, patriotism, and empire: Lessons for England, Spain, and the New World in the works of Richard Hakluyt', *Renaissance Quarterly* LII/3 (fall, 2009), pp. 809–58.

Bosher, J.F., 'Vancouver Island in the Empire', *Journal of Imperial and Commonwealth History* XXXIII/3 (September 2005), pp. 349–68.

Browne, Janet, 'A science of empire: British biogeography before Darwin', *Revue d'histoire des sciences* XLV/4 (1992), pp. 453–75.

Buckner, Phillip, 'The royal tour of 1901 and the construction of an imperial identity in South Africa', *South African Historical Journal* XLI/1 (1999), pp. 324–48.

Cain, Peter J., 'Empire and the languages of character and virtue in later Victorian and Edwardian Britain', *Modern Intellectual History* IV/2 (August, 2007), pp. 249–73.

Cassell, Frank A., 'Slaves of the Chesapeake Bay area and the war of 1812', *The Journal of Negro History* LVII/2 (April, 1972), pp. 144–55.

Clayton, Daniel, 'The creation of imperial space in the Pacific Northwest', *Journal of Historical Geography* XXVI/3 (2000), pp. 327–50.

Cohen, Andrew, 'Lonrho and oil sanctions against Rhodesia in the 1960s', *Journal of Southern African Studies* XXXVII/4 (2011), pp. 715–30.

Cormack, Lesley B., 'Britannia rules the waves?: images of empire in Elizabethan England', in Andrew Gordon and Bernhard Klein (eds), *Literature, Mapping and the Politics of Space in Early Modern Britain* (Cambridge, 2001) , pp. 45–68.

Connelly, Mark, 'Battleships and British society, 1920–1960', *International Journal of Naval History* III/2–3 (August/December, 2004).

Dahl, Erik J., 'Naval innovation: From coal to oil', *Joint Force Quarterly*, 27 (January, 2001), pp. 50–6.

Duder, C.J., ' "Men of the officer class": The participants in the 1919 soldier settlement scheme in Kenya', *African Affairs* XCII/366 (January, 1993), pp. 69–87.

Duffy, Michael, 'The establishment of the Western Squadron as the linchpin of British naval strategy', in M. Duffy (ed.), *Parameters of British Naval Power* (Exeter, 1992).

Ehrensaft, Philip, 'The political economy of informal empire in pre-colonial Nigeria, 1807–1884', *Canadian Journal of African Studies* VI/3 (1972), pp. 451–90.

Enloe, Cynthia H., 'Ethnicity in the evolution of Asia's armed bureaucracies', in C. DeWitt Ellinwood and Cynthia H. Enloe (eds), *Ethnicity and the Military in Asia* (London, 1981), pp. 1–17.

Ewald, Janet J., 'Crossers of the sea: slaves, freedmen, and other migrants in the northwestern Indian Ocean, *c*.1750–1914', *American Historical Review* CV/1 (February, 2000), pp. 69–91.

Fair, Laura, ' "It's just no fun anymore": Women's experiences of Taarab before and after the 1964 Zanzibar Revolution', *The International Journal of African Historical Studies* XXXV/1 (2002), pp. 61–81.

Foy, Charles R., 'Uncovering hidden lives', *Common-Place* IX/ 2 (January, 2009).

Fulford, Tim, 'Romanticism, the South seas and the Caribbean: the fruits of empire'. *European Romantic Review* XI/4 (2000), pp. 408–34.

Gallagher, John and Ronald Robinson, 'The imperialism of free trade', *Economic History Review* VI/1 (1953), pp. 1–15.

Gimblett, Richard H., 'Reassessing the Dreadnought Crisis of 1909 and the origins of the Royal Canadian Navy', *The Northern Mariner* IV/1 (January, 1994), pp. 35–53.

Gordon, Donald C., 'The Admiralty and dominion navies, 1902–1914', *The Journal of Modern History* XXXIII/4 (December, 1961), pp. 407–22.

Gough, Barry, M., 'Send a gunboat! Checking slavery and controlling liquor traffic among coast Indians of British Columbia in the 1860s', *The Pacific Northwest Quarterly* LXIX/4 (October, 1978), pp. 159–68.

——'The British reoccupation and colonization of the Falkland Islands, or Malvinas, 1832–1843', *Albion* XXII/2 (summer, 1990), pp. 261–87.

Graham, Gerald S., 'The naval defence of British North America 1739–1763', *Transactions of the Royal Historical Society* xxx (December, 1948), pp. 95–110.

Grant, John N., 'Black immigrants into Nova Scotia, 1776–1815', *The Journal of Negro History* LVIII/3 (July, 1973), pp. 253–70.

Harrington, Ralph, ' "The mighty hood": Navy, empire, war at sea and the British national imagination, 1920–60', *Journal of Contemporary History* XXXVIII/2 (April, 2003), pp. 171–85.

Headrick, Daniel R., 'The tools of imperialism: Technology and the expansion of European colonial empires in the nineteenth century', *The Journal of Modern History* LI/2 (June 1979), pp. 231–63.

Hill, Errol G., 'Calypso and War', *Black American Literature Forum* XXIII/1 (spring, 1989).

Hopper, Matthew S., 'East Africa and the end of the Indian Ocean slave trade', *Journal of African Development* XIII/1 (spring, 2011), pp. 39–65.

Hurd, Archibald, 'The racial war in the Pacific: an imperial peril', *Fortnightly Review*, XCIII/558 (June, 1913).

Jack, Marian, 'The purchase of the British government's shares in the British Petroleum Company 1912–1914', *Past and Present* XXXIX (April, 1968), pp. 139–68.

Jackson, Ashley, 'The Royal Navy and the Indian ocean region since 1945', *The RUSI Journal* CLI/6 (2006), pp. 78–82.

Johnson, Howard, 'Oil, imperial policy and the Trinidad disturbances, 1937', *Journal of Imperial and Commonwealth History* IV/1 (1975), pp. 29–54.

Jones, G.G., 'The state and economic development in India 1890–1947: the case of oil', *Modern Asian Studies* XIII/3 (1979), pp. 353–75.

Jones, Leigh and Nicky Jones, 'Emergence, cessation and resurgence during the evolution of Rugby Union in Hong Kong', *The International Journal of the History of Sport* XXIX/9 (2012), pp. 1344–62.

Keate, George, 'Verses to Captain Bligh', in James Stanier Clarke, John Jones, and Stephen Jones (eds), *The Naval Chronicle*, Vol. 4 (London, 1801).

Killingray, David, 'The idea of a British imperial African army', *Journal of African History*, xx/3 (1979), pp. 421–36.

Knellwolf, Christa, 'The exotic frontier of the imperial imagination', *Eighteenth-Century Life* XXVI/3 (fall, 2002), pp. 10–30.

Lambert, Andrew, 'Economic power, technological advantage, and imperial strength: Britain as a unique global power, 1860–1890', *International Journal of Naval History* v/2 (August, 2006).

Law, Alex, 'Of navies and navels: Britain as a mental islands', *Geografiska Annaler: Series B, Human Geography* LXXXVII/4 (December, 2005), pp. 267–77.

Lincoln, Margarette, 'Mutinous behaviour on voyages to the South Seas and its impact on eighteenth-century civil society', *Eighteenth-Century Life* XXXI/1 (2007), pp. 62–80.

Low, D.A. and J. Lonsdale, 'Introduction', in D.A. Low and Alison Smith (eds), *The Oxford History of East Africa* (Oxford, 1976).

Lynch, John, 'British policy and Spanish America, 1783–1808', *Journal of Latin American Studies* I/1 (May, 1969), pp. 1–30.

MacDonald, Fraser, 'The last outpost of empire: Rockall and the Cold War', *Journal of Historical Geography* XXXII (2006), pp. 627–47.

MacRae, Christopher and Tony Laurence, 'The 1964 Tanganyika Rifles mutiny and the British armed intervention that ended it', *The RUSI Journal* CLII/2 (2007), pp. 96–101.

McCreery, Cindy, 'Telling the story: HMS Galatea's voyage to South Africa, 1867', *South African Historical Journal* LXI/4 (2009), pp. 817–37.

McIntyre, David W., 'New Zealand and the Singapore base between the wars', *Journal of Southeast Asian Studies* II/1 (March, 1971), pp. 2–21.

Mclean, David, 'Trade, politics and the navy in Latin America: the British in the Parana, 1845–46', *Journal of Imperial and Commonwealth History* xxxv/3 (August 2007), pp. 351–70.

Millett, Nathaniel, 'Britain's 1814 occupation of Pensacola and America's response: An episode of the war of 1812 in the southeastern borderlands', *The Florida Historical Quarterly* LXXXIV/2 (fall, 2005), pp. 229–55.

Mitcham, John C., 'Navalism and Greater Britain, 1897–1914', in Duncan Redford (ed.), *Maritime History and Identity: The Sea and Culture in the Modern World* (London, 2013), pp. 271–93.

Mobley, Richard, 'The Beira patrol: Britain's broken blockade against Rhodesia', *Naval War College Review* LV/1 (winter, 2000), pp. 63–84.

Mulchansingh, Vernon C., 'The oil industry in the economy of Trinidad', *Caribbean Studies* XI/1 (April, 1971), pp. 73–100.

Murphy, Elaine, 'The navy and the Cromwellian conquest of Ireland, 1649–53', *Journal of Maritime Research* XIV/1 (2012), pp. 1–13.

Nicholls, Bob, 'Colonial naval forces before Federation', in David Stevens and John Reeve (eds), *Southern Trident: Strategy, History and the Rise of Australian Naval Power* (Crows Nest, 2001), pp. 125–39.

Nish, Ian, 'Japan in Britain's view of the international system, 1919–37', in Ian Nish (ed.), *Anglo-Japanese Alienation, 1919–1952: Papers of the Anglo-Japanese Conference on the History of the Second World War* (Cambridge, 1982).

O'Brien, Phillips Payson, 'The titan refreshed: Imperial overstretch and the British navy before the First World War', *Past and Present*, 172 (August, 2001), pp. 146–69.

Offer, Avner, 'The British Empire, 1870–1914: A waste of money?', *Economic History Review* XLVI/2 (1993), pp. 215–38.

Owen, Janet, 'Collecting artefacts, acquiring empire: Exploring the relationship between Enlightenment and Darwinist collecting and late-nineteenth-century British imperialism', *Journal of the History of Collections* XVIII/1 (2006), pp. 9–25.

Owens, Geoffrey Ross, 'Exploring the articulation of governmentality and sovereignty: The Chwaka Road and the bombardment of Zanzibar, 1895–1896', *Journal of Colonialism and Colonial History* VIII/2 (2007).

Pechey, Graham, ' "Empire laid up in heaven": Postcoloniality and eternity', *Critical Quarterly* XLIX/2 (summer, 2007), pp. 1–30.

Pereira, Clifford, 'Black liberators: The royal of Africans and Arab sailors in the Royal Navy within the Indian Ocean 1841–1941', research paper presented at the UNESCO symposium in Rabat, Morocco (May, 2007).

Pincus, Steven C.A., 'Popery, trade and universal monarchy: The ideological context of the outbreak of the second Anglo-Dutch war', *The English Historical Review* CVII/422 (January, 1992), pp. 1–29.

Pollak, Oliver B., 'The origins of the second Anglo-Burmese war (1852–53)', *Modern Asian Studies* XII/3 (1978), pp. 483–502.

Potgieter, T.D., 'Maritime defence and the South African Navy to the cancellation of the Simon's Town Agreement', *Scientia Militaria: South African Journal of Military Studies* xxx/2 (2000), pp. 159–82.

Pratt, Lynda, 'Naval contemplation', *Journal for Maritime Research* II/1 (2000), pp. 84–105.

Raj, Kapil, '18th-Century pacific voyages of discovery, "big science", and the shaping of a European scientific and technological culture', *History and Technology: An International Journal* XVII/2 (2000), pp. 79–98.

Ransom, Bernard, 'A nursery of fighting seamen? The Newfoundland Royal Naval Reserve, 1901–1920', in Michael L. Hadley, Rob Huebert and Fred W. Crickard (eds), *Nation's Navy: In Quest of Canadian Naval Identity*, pp. 239–252.

Rayner, Jonathan, ' "Entrusted with the ruling of the waves": Images of the post-war royal navy in the NMM film archive', *Journal of Maritime Research* x/1 (2008), pp. 50–66.

Redford, Duncan, 'The "hallmark of a first-class navy" The nuclear-powered submarine in the Royal Navy 1960–77', *Contemporary British History* XXIII/2 (May, 2009), pp. 181–97.

Reinwald, Brigitte, ' "Tonight at the empire": Cinema and urbanity in Zanzibar, 1920s to 1960s', *Verdier: Afrique & histoire* v/1 (2006), pp. 81–109.

Rodger, N.A.M., 'The Drake-Norris expedition: English naval strategy in the sixteenth century', *Militaria: Revista de Cultura Militar* 8 (1996), pp. 89–98.

Rohlehr, Gordon, 'I lawa: the construction of masculinity in Trinidad and Tobago Calypso', in Rhoda E. Reddock, *Interrogating Caribbean Masculinities: Theoretical and Empirical Analyses* (University Press of the West Indies, 2004).

Sacks, David Harris, 'The true temper of empire: Dominion, friendship and exchange in the English Atlantic, c.1575–1625', *Renaissance Studies* XXVI/4 (September, 2012), pp. 531–58.

Samson, Jane, ' "That extensive enterprise": HMS *Herald*'s North Pacific survey, 1845–1851', *Pacific Science* LII/4 (1998), pp. 287–93.

Sarty, Roger, 'Canadian maritime defence, 1892–1914', *Canadian Historical Review* LXXI/4 (1990), pp. 462–90.

Sea Power Centre, 'The Great White Fleet's 1908 visit to Australia', *Semaphore*, 8 (July, 2008).

Sherman, William H., 'Putting the British seas on the map: John Dee's imperial cartography', *Cartographica* xxxv/3/4 (autumn/winter, 1998), pp. 1–10.

Sherwood, Marika, 'More on blacks in the Royal Navy', *Black and Asian Studies Association Newsletter* 26 (January, 2000), pp. 8–10.

Sorrenson, Richard, 'The ship as a scientific instrument in the eighteenth century', *Osiris*, 2nd Series XI (1996), pp. 221–36.

Speller, Ian, 'An African Cuba? Britain and the Zanzibar Revolution, 1964', *Journal of Imperial and Commonwealth History*, Vol. 35, No. 2 (June 2007), pp. 283-301.

Speller, Ian, 'The seaborne/airborne concept: littoral manoeuvre in the 1960s?', *Journal of Strategic Studies* XXIX/1 (2006), pp. 53–82.

Spence, Daniel Owen, 'Australian naval defence and the 1887 colonial conference: context, policy, and reaction', *International Journal of Naval History* VI/1 (April, 2007).

——'Imperial transition, indianisation and race: developing national navies in the subcontinent, 1947–1964', *South Asia: Journal of South Asian Studies* XXXVII/2 (June, 2014), pp. 323–38.

Stafford, Robert A., 'Geological surveys, mineral discoveries, and British expansion, 1835–71', *Journal of Imperial and Commonwealth History*, Vol. 12, No. 3 (1984), pp. 5–32.

Stockey, Gareth, 'Sport in Gibraltar – problematizing a supposed "problem", 1713–1954', *Sport in History* XXXII/1 (2012), pp. 1–25.

Strobel, Margaret, 'Women's wedding celebrations in Mombasa, Kenya', *African Studies Review* XVIII/3, Women in Africa (December, 1975), pp. 35–45.

Subritzky, John, 'Britain, *Konfrontasi*, and the end of empire in Southeast Asia, 1961–65', *Journal of Imperial and Commonwealth History* XXVIII/3 (2000), pp. 209–27.

Tsuruta, Tadasu, 'Popular music, sports, and politics: A development of urban cultural movements in Dar es Salaam, 1930s–1960s', *African Study Monographs* XXIV/3 (2003), pp. 195–222.

Upton, L.F.S., 'Contact and conflict in the Atlantic and Pacific coasts of Canada', *Acadiensis* IX/2 (spring, 1980), pp. 3–13.

Van der Linden, Marcel, 'Unanticipated consequences of "humanitarian intervention": the British campaign to abolish the slave trade, 1807–1900', *Theor. Soc.* XXXIX (2010), pp. 281–98.

Wilde, Richard H., 'The Boxer affair and Australian responsibility for imperial defense', *Pacific Historical Journal* XXVI/1 (February, 1957), pp. 51–65.

Williams, Glyndwr, ' "To make discoveries of countries hitherto unknown'; the Admiralty and Pacific exploration in the eighteenth century', in Alan Frost and Jane Samson (eds), *Pacific Empires: Essays in Honour of Glyndwr Williams* (Melbourne, 1999).

Withers, F.M., 'Nyasaland in 1895–96', *Nyasaland Journal* II/1 (January, 1949), pp. 16–34.

Wolff, Richard D., 'British imperialism and the East African slave trade', *Science and Society* XXXVI/4 (winter, 1972), pp. 443–62.

# Books

Baines, Andrew, *A History of the Royal Navy: The Victorian Age* (London, 2014).

——*A History of the Royal Navy: The Age of Sail* (to be published 2016).

Bingham, Commander J. Elliot, RN, *Narrative of the Expedition to China, from the Commencement of the War to its Termination in 1842; With Sketches of the Manners and Customs of that Singular and Hitherto Almost Unknown Country*, Vol. 1 (London, 1843).

Black, Jeremy, *The British Seaborne Empire* (London, 2004).

Brooks, George E. Jr, *The Kru Mariner in the Nineteenth Century: An Historical Compendium*, Liberian Studies Monograph Series No. 1 (Newark, 1972).

Brooks, Richard, *The Long Arm of Empire: Naval Brigades from the Crimea to the Boxer Rebellion* (London, 1999).

Cellier, Francois and Cunningham Bridgeman, *Gilbert and Sullivan and Their Operas* (London, 1914).

Churchill, Winston, *The Grand Alliance* (London, 1950).

Clowes, Wm. Laird, *The Royal Navy: A History from the Earliest Times to the Death of Queen Victoria*, Vol. 7 (London, 1903).

Conley, Mary, *From Jack Tar to Union Jack: Representing Naval Manhood in the British Empire, 1870–1918* (Manchester, 2009).

Cormack, Lesley B., *Charting an Empire: Geography at the English Universities 1580–1620* (Chicago, 1997).

Crane, David, *Scott of the Antarctic: A Life of Courage, and Tragedy in the Extreme South* (London, 2005).

Craton, Michael, *Founded Upon the Seas: A History of the Cayman Islands and Their People* (Kingston, 2003).

Dampier, William, *A Voyage to New Holland, Etc., In the Year 1699* (London, 1729).

Davis, Ronald W., *Ethnohistorical Studies on the Kru Coast*, Liberian Studies Monograph Series No. 5 (Newark, 1976).

Deacon, Richard, *John Dee* (London, 1968).

de Silva, Jayasuriya Shihan and Richard Pankhurst, *The African Diaspora in the Indian Ocean* (Trenton, 2003).

Dingman, Roger, *Power in the Pacific: The Origins of Naval Arms Limitation, 1914–1922* (Chicago, 1976).

Elkins, Charles, *The Naval Battles of Great Britain: From the Accession of the Illustrious House of Hanover to the Throne to the Battle of Navarin* (London, 1828).

Endicott, Stephen Lyon, *Diplomacy and Enterprise: British China Policy, 1933–1937* (Manchester, 1975).

Evelyn, John, *Navigation and Commerce, the Origins and Progress* (London, 1674).

Fain, W. Taylor, *American Ascendance and the British Retreat in the Persian Gulf Region* (Palgrave, 2008).

Ferraby, H.C., *The Imperial British Navy: How the Colonies Began to Think Imperially Upon the Future of the Navy* (London, 1918).

Field, Andrew, *Royal Navy Strategy in the Far East, 1919–1939: Preparing for War against Japan* (London, 2004).

Fisher, Herbert A.L., *The History of England from the Accession of Henry VII to the Death of Henry VIII, 1485–1547* (London, 1910).

Flourney, Francis Rosebro, *British Policy towards Morocco in the Age of Palmerston* (London, 1935).

Frost, Diane, *Work and Community among West African Migrant Workers since the Nineteenth Century* (Liverpool, 1999).

Ghazemy, Mahmud and Mohd. Haidar Abu Bakar (eds.), *Royal Malaysian Navy: 55 Years* (Kuala Lumpur, 1990).

Goodwin, R.B., *Hongkong Escape* (London, 1953).

Grant, J. (ed.), 'The Old Scots Navy from 1689 to 1710', *The Navy Records Society*, Vol. 44 (London, 1914).

Haggie, Paul, *Britannia At Bay: The Defence of the British Empire against Japan, 1931– 1941* (Oxford, 1981).

Hall, Commander William Hutcheon, RN and William Dallas Bernard, *Narrative of the Voyages and Services of the Nemesis from 1840 to 1843; and of the Combines Naval and Military Operations in China*, 2nd ed. (London, 1845).

Hermann, Arthur, *To Rule the Waves: How the British Navy Shaped the Modern World* (Chatham, 2005).

Holland, Robert, *Blue-Water Empire: The British in the Mediterranean since 1800* (London, 2012).

Howell, Raymond C., *The Royal Navy and the Slave Trade* (London, 1987).

Ingrams, W.H., *Zanzibar: Its History and Its People* (London, 1967).

Jackson, Ashley, *War and Empire in Mauritius and the Indian Ocean* (Basingstoke, 2001).

——*The British Empire and the Second World War* (London, 2006).

Jebb, Richard, *The Imperial Conference*, Vol. I (London, 1911).

Jose, Arthur W., *The Royal Australian Navy* (Queensland, 1987).

Kelshall, Gaylord, *The U-Boat War in the Caribbean* (Annapolis, 1994).

Kennedy, Paul, *The Rise and Fall of British Naval Mastery* (London, 1976).

Langford, Paul, *Short Oxford History of The British Isles: The Eighteenth Century:1688– 1815* (Oxford, 2002).

Lavery, Brian, *Shield of Empire: The Royal Navy and Scotland* (Edinburgh, 2007).

Luard, Tim, *Escape from Hong Kong: Admiral Chan Chak's Christmas Day Dash, 1941* (Hong Kong, 2012).

Lucas, Sir Charles (ed.), *The Empire At War*, Vol. I (Oxford, 1921).

Mackinder, Harold, *Britain and the British Seas* (London, 1902).

Mahan, Alfred Thayer, *The Influence of Sea Power Upon History, 1660–1783*, 12th ed. (Boston, 1890/1918).

Malone, Edmund (ed.), *The Critical and Miscellaneous Prose Works of John Dryden*, Vol. 1, Part 2 (London, 1800).

Mann, Kristin, *Slavery and the Birth of an African City: Lagos, 1760–1900* (Indiana, 2007).

Marder, Arthur J., *Old Friends, New Enemies: The Royal Navy and the Imperial Japanese Navy, Strategic Illusions, 1936–1941* (Oxford, 1981).

Marshall, Oliver (ed.), *The Caribbean at War, British West Indians in World War II* (London, 1992).

McLaughlin, Norman Rudolph, *The Forgotten Men of the Navy* (Miami, 2002).

Ngũgĩ wa Thiong'o, *Decolonising the Mind: The Politics of Language in African literature* (London, 1986).

Nordin, Dato Yusof (ed.), *Honour and Sacrifice* (Kuala Lumpur, 1994).

O'Connor, V.C. Scott, *The Empire Cruise* (London, 1925).

Patience, Kevin, *Zanzibar, Slavery and the Royal Navy* (Bahrain, 2000).

Postlethwayt, Malachy, *The Universal Dictionary of Trade and Commerce, with Large Additions and Improvements*, Vol. 2 (London, 1774).

Potter, Simon J., *News and the British World: The Emergence of an Imperial Press System* (Oxford, 2003).

Redford, Duncan, *The Submarine: A Cultural History from the Great War to Nuclear Combat* (London, 2010).

Robinson, C. and H.M. Ross (eds), *Brassey's Naval and Shipping Annual* (London, 1931).

Robson, Martin, *A History of the Royal Navy: The Napoleonic Wars* (London, 2014).

Rodger, N.A.M., *The Wooden World: An Anatomy of the Georgian Navy* (New York, 1986).

Rosberg, Carl G. Jr and John Nottingham, *The Myth of 'Mau Mau': Nationalism in Kenya* (London, 1966).

Royal Malaysian Navy (ed.), *Serving the Nation* (Kuala Lumpur, 2004).

Rüger, Jan, *The Great Naval Game: Britain and Germany in the Age of Empire* (Cambridge, 2007).

Samson, Jane, *Imperial Benevolence: Making British Authority in the Pacific Islands* (Hawaii, 1998).

Schull, Joseph, *Far Distant Ships: An Official Account of Canadian Naval Operations in World War II* (Toronto, 1987).

Selwyn-Clarke, Sir Selwyn, *Footprints: The Memoirs of Sir Selwyn Selwyn-Clarke* (Hong Kong, 1975).

Simpson, Donald, *Dark Companions: The African Contribution to the European Exploration of East Africa* (London, 1975).

Singh, Satyindra, *Blueprint to Bluewater: The Indian Navy 1951–65* (New Delhi, 1992).

Smith, Robert Sydney, *The Lagos Consulate, 1851–1861* (Berkeley, 1979).

Snow, Phillip, *The Fall of Hong Kong: Britain, China and the Japanese Occupation* (London, 2003).

Spence, Daniel Owen, *Colonial naval culture and British imperialism, 1922-67* (Manchester, 2015).

Spence, Jonathan D., *Chinese Roundabout: Essays in History and Culture* (London, 1992).

Tracy, Nicholas (ed.), *The Collective Naval Defence of the Empire 1900–1944* (Aldershot, 1997).

Tsang, Steve, *A Modern History of Hong Kong* (London, 2007).

Wise, Jon, *The Role of the Royal Navy in South America, 1920–1970* (London, 2014).

## Newspapers

*South China Morning Post*
*The Straits Times*
*The Telegraph*
*The Times*

## Oral Interviews

Ahmad, Dahim bin Noordin, conducted by Daniel Chew on 5 March 1992, 001318, reel 3, National Archives of Singapore.

Karu Selvaratnam, conducted by the author on 30 June 2009 in Kuala Lumpur.

Kuttan, Chitharanjan, conducted by Jason Lim in September 2002, 002697, reel 4, National Archives of Singapore.

McCoy, Harry, conducted by Heather McLoughlin on 3 July 1991, tape 2A, p. 10, Cayman Islands National Archive.

Tan, Reverend Neivelle, conducted by the author on 14 July 2009 in Singapore.

Thanabalasingam, Tan Sri, conducted by the author on 25 June 2009 in Kuala Lumpur.

## Unpublished Manuscripts

Baptiste, Fitzroy A., 'The European possessions in the Caribbean in World War II: dimensions of great power co-operation and conflict', Ph.D. thesis, University of the West Indies, Trinidad (1981).

Fosten, Peter, 'A Far East Odyssey, part two: aiding the civil power, Malaya 1950–53', personal memoir acquired from the author in Newton Abbot on 14 August, 2009.

Gray, Steven, 'Black diamonds: coal, the Royal Navy, and British imperial coaling stations, *circa* 1870–1914', Ph.D. thesis, University of Warwick (March, 2014).

Swain, Derek Anthony, 'A history of sport in British Columbia to 1885: a chronicle of significant developments and events', Master's thesis, University of British Columbia (April, 1977).

# Index

Page references in *italics* refer to illustrations.